JAPANESE STAMP SPECIALIZED CATALOGUE

Commemoratives 1952-1966

JSCA

Japanese/English

日本切手専門カタログ

ビジュアル日専

記念・特殊切手編 1952-1966

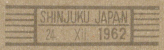

JAPANESE STAMP SPECIALIZED CATALOGUE

Japanese/English

ビジュアル日専
記念・特殊切手編 1952-1966
Commemoratives 1952-1966

日本切手専門カタログ

JSCA

目次 CONTENTS

ビジュアル日専の使い方　　How to use this catalogue "JSCA" ……………………………… 4

記念・特殊切手　Commemorative and Special Stamps 1952-1966 ……… 8
- (1) 額面「銭位」時期　　"Sen" Value Era 1952 ……………………………………………… 14
- (2) 額面「円位」時期、ローマ字なし　"Yen" Value Era, without NIPPON 1952-1965 ……… 14
- (3) 額面「円位」時期、ローマ字入り　"Yen" Value Era, with NIPPON 1966 ……………… 95
- 使用例評価　　Valuations by Cover ……………………………………………………… 99
- 消印別評価　　Valuations by Postmark ………………………………………………… 102

公園切手　National Park Stamps 1958-1966 ……………………………………… 104
- I. 第2次国立公園切手　2nd National Park Series 1962-1965 ………………………… 106
- II. 国定公園切手　Quasi-National Park Series 1958-1966 …………………………… 119

年賀切手　New Year's Greeting Stamps 1952-1966 …………………………… 132
- 使用例評価　　Valuations by Cover ……………………………………………………… 143
- 消印別評価　　Valuations by Postmark ………………………………………………… 144

ステーショナリー Postal Stationery

記念・特殊はがき	Commemorative and Special Postal Cards	145
年賀はがき	New year's Greeting Postal Cards	146
季節見舞はがき	Seasonal Greeting Postal Cards	149

巻末資料 Appendices

製版・目打組合せ型式	Types of Plate and Perforation Configuration	150
ご当地局の定義と類型	Definition and Types of Related Area Post Office	155
郵便料金変遷表	Transition of Postage Rates	161
西暦対照表	Comparative List of Japanese, Chinese, Manchuria and Gregorian Calenders	164

●本カタログの採録範囲 Stamps included in this catalogue

記念・特殊切手 Commemorative and Special Stamps	C225-C445, C453, C456-C457
第2次国立公園切手 2nd National Park Series Stamps	P91-P118
国定公園切手 Quasi-National Park Series Stamps	P200-P225
年賀切手 New Year's Greeting Stamps	N7-N21

広告もくじ（社名50音順）

宇中スタンプ	168	タカハシスタンプ商会	裏表紙
英国海外郵趣代理部	168	日本フィラテリックセンター	168
越中趣味の会	167	日本郵趣協会	167
小林スタンプ商会	168	モナリザスタンプ	167
セキネ・スタンプ	167	郵趣サービス社	表紙裏・裏表紙裏
ゼネラルスタンプ	167	ロータスフィラテリックセンター	168

ビジュアル日専の使い方
How to use this catalogue "JSCA"

1.『ビジュアル日専』の特徴

本書『ビジュアル日専（日本切手専門カタログ）』は、日本で最も発行部数が多い普及版『さくら日本切手カタログ』、国際切手展の文献クラスで金賞を受賞した『日本普通切手専門カタログ』と同様に、公益財団法人日本郵趣協会が監修しています。
ビジュアル日専はこれまでの切手カタログを発展させたもので、その特徴は次の通りです。
（1）切手の色調を明確に分類するため、カラー図版を用いてビジュアルで示しています。
（2）日本切手の学術研究を発展させるため、日本語と英語の2カ国語表記を用いています。
（3）専門的評価手法を実施するため、用紙、色調、目打、使用例、消印別の評価を示しています。

2.カタログ評価

郵便切手類の取引形態には、大きく分類して次の3つのタイプがあります。
（1）切手商の店頭または通信販売による取引
（2）競売（オークション）方式による取引
（3）切手商対個人、個人対個人による取引
ビジュアル日専は、（1）～（3）を総合的に勘案した評価を示していますが、取引形態によりその価格は大きく異なる場合があります。カタログ評価は、郵便切手類を購入する場合の目安となる金額ですが、郵便切手類を売却する場合にはカタログ評価を大きく下回ることもあります。また本書のカタログ評価は、オープン取引、プライベート取引を拘束するものではありません。

3.切手データ
❶切手名
切手名は必ずしも報道発表名でなく、愛称となっている切手名を示す場合があります。
❷図版
切手単片、ステーショナリー印面は実物の75%～100%、切手帳は50%、カバー類は35%～45%で示しています。
❸特印、風景印
初日カバーに押された特印、風景印の初日印の適応局名を示します。
❹発行日
公示された発行日で、西暦で年、月、日の順で示します。

1.Features of "JSCA"

This catalogue ("JSCA") is supervised by the Japan Philatelic Society, as are both the "Sakura Catalogue of Japanese Stamps", which has the largest circulation in Japan, and the "Japanese Stamp Specialized Catalogue", which won the gold medal in the philatelic literature class at the International Stamp Exhibition.
JSCA is a new variant on the conventional stamp catalogue. Its features are as follows.
（1）To classify the color shade of stamps clearly, the JSCA uses color illustrations.
（2）To foster academic research on Japanese stamps globally, the text of JSCA is written in both English and Japanese.
（3）To implement specialized evaluation methods, the JSCA shows the value based on papers, color shades, perforations, covers, and postmarks.

2.Catalogue Values

There are three major types of transaction in philatelic material, such as postage stamps, covers, postal stationery, and the like:
（1）Transactions at stamp dealers' premises, or by mail order;
（2）Transactions by philatelic auction;
（3）Transactions between stamp dealers and individuals, and between individuals.
JSCA shows the catalogue values with all the factors through (1) to (3) taken into consideration. However, actual values may vary greatly depending on the transaction type. The catalogue value is a standard amount when purchasing philatelic materials, but values realized when selling such materials may fall significantly below the catalogue value. In addition, the catalogue values published in JSCA do not bind any open transactions or private transactions.

3.Basic Information
❶Names
The name of the stamp is not necessarily the name in the official press release, but may indicate a nickname for the stamp.
❷Illusration
Stamps and postal stationery are illustrated at 75% to 100% of the actual size, booklet panes at 50%, and covers at 35% to 45%.

記念・特殊切手の例 Commemorative and Special Stamps

❶ **C427** 国土緑化運動　❹発行日：1965（昭和40）.5.9
❺　National Land Afforestation Campaign　Date of Issue : 1965 (Showa 40) .5.9

❸●特印
Special Date Stamp

米子 Yonago

❺ C427 10 yen 多色 multicolored ⋯⋯⋯⋯ ** 50　● 30　FDC 200
　　❺　❻　　　❼　　　　　　❽　　　❾

切手データ Basic Information	
❿ 版式：グラビア輪転版	printing plate : rotary photogravure plate
⓫ 目打：櫛型13½	perforation : comb perf. 13½
⓬ 印面寸法：27×31mm	printing area size : 27×31mm
⓭ 発行枚数：2,400万枚	number issued : 24 million
⓮ シート構成：20枚（横4×縦5）	pane format : 20 subjects of 4×5
⓯ 製版・目打組合せ：タイプ1C-5	plate and perforation configuration : type 1C-5

❺ C427 樹木と陽光 Stylized Tree and Sun

❺ **C427** ●定常変種 Constant Flaws

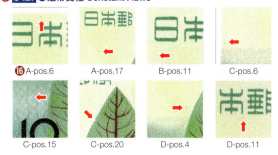

位置 position	特徴 feature	**	●
A-pos.6	"本"の上に黒点 black dot above "本"	200	100
A-pos.17	"日"の右下方に黒点 black dot below right of "日"		
B-pos.11	"日"の左下に緑点 green dot under left of "日"		
C-pos.6	左マージン付近に大小2個の黒点 two large and small black dots near left margin		
C-pos.15	"10"の上方に白抜け white space upper "10"		
C-pos.20	樹木の線欠け missing line in stylized tree		
D-pos.4	樹木の左に黒点2個と赤点 two black dots and red dot to left of stylized tree		
D-pos.11	"本"の右下に赤点 red dot under right of "本"		

⓰ ❽

（ ）内は、和暦を示します。公示のない場合には最初期使用日を示し、発行日の詳細が確定できない場合には「－」とします。

❺ **カタログ番号**
カタログ番号は、『さくら日本切手カタログ』、『日本普通切手専門カタログ』と共通です。カタログ番号の英記号は、p.7の別表を参照ください。

❻ **額面**
通貨は円、銭の2種類です。
通貨単位は1円＝100銭です。

❼ **刷色**
色名は、『日本切手カラーガイド』、『スコット』カタログの色名表記を基本としています。

❽ **評価**
記号の略称は次の通りで、評価額は円で示します。
**：未使用極美品
●：使用済
評価額には、次の記号を使用する場合があります。
u：実存の可能性はあるが未発見
x：実存しない
－：残存数が極度に少なく、取引実績がなく評価が未確定

❸ **Special Date Stamps and Pictorial Postmarks**
This indicates the name of the post office that was designated for the first day cancellation using a special date stamp or pictorial postmark.

❹ **Date of Issue**
The date of issue is the official first day of issue and is shown in the Gregorian calendar in year-month-day order. The date shown in (　) is in the Japanese (regnal year) calendar.

❺ **Catalogue Number**
The catalogue numbers in JSCA are the same as the catalogue numbers of the "Sakura Catalogue" and the "Japanese Stamp Specialized Catalogue". The catalogue number prefixes and classifications are listed on page 7.

❻ **Face Value**
There are two currency units: yen and sen. One yen equals 100 sen.

❼ **Color Shade**
The names used for colors are based on the "Japanese Stamp Color Shade Guide (Japan Philatelic Society)" and the "STANLEY GIBBONS STAMP COLOUR KEY".

❽ **Catalogue Values**
The abbreviations for symbols are as follows, and the values are shown in yen.

公園切手の例 National Park Stamps

❶ P218 玄海国定公園 ❹発行日：1963（昭和38）.3.15
❺ **Genkai Quasi-National Park** Date of Issue：1963 (Showa 38) .3.15
❶

●風景印 Pictorial Postmarks

❺ ❻ ❼ ❽ ❾
P218 10 yen 多色 multicolored ⋯⋯⋯⋯⋯⋯⋯ 60 40 280
　　　　　　　　　　　　　　　　　　　　 ＊＊　◉ FDC

P218
芥屋の大門
Keya-no-Oto
Rock

芥屋 Keya

切手データ Basic Information	
❸発行枚数：1,000万枚	number issued：10 million
❻製版・目打組合せ：タイプ 2B-2	plate and perforation configuration：type 2B-2

●変則目打 Irregular Perforation

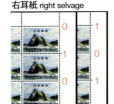

右耳紙 right selvage

図1 fig.1　図2 fig.2

1) この切手の目打穿孔は連続櫛型目打の二連1型である。目打の変種として一部のシートに櫛歯中段の横目打右側の目打針が1本少なく植えられているものが存在する。

1) The perforation of this stamp was the continuous double comb perforation with one extension hole, of which the right extension hole in the middle row was omitted.

図1 fig.1 目打穴数 number of extension hole 0-1-0-1-0
図2 fig.2 目打穴数 number of extension hole 1-0-1-0-1

❺ **P218** ●定常変種 Constant Flaws

❻ A-pos.8　B-pos.9　C-pos.19　D-pos.5

位置 position	特徴 feature	＊＊	◉
A-pos.8	"郵"の右に青点 blue dot to right of "郵"	200	100
B-pos.9	"郵"と"便"の間に小青点 small blue dot between "郵" and "便"		
C-pos.19	"0"の右に茶点 brown dot to right of "0"		
D-pos.5	岩の上方に青点2ヵ所 two blue dots above rock		

❾初日カバー評価
初日カバーの評価額は円で示します。

❿版式
印刷方法の種類を示します。

⓫目打
目打の型式を示します。

⓬印面寸法
切手の図案、額面が印刷されている寸法で、切手全体の寸法ではありません。大きさは横×縦で示し、単位はミリです。

⓭発行枚数
1種あたりの発行枚数を示します。小型シート、切手帳ペーンの場合は、部数を示します。

⓮シート構成
1シートの切手枚数で、（　）内は横×縦の切手枚数を示します。

⓯製版・目打組合せ
本カタログでは、切手1種ずつの実用版構成、目打穿孔、目打の抜け方の組み合わせをタイプ別に示しています。詳細は、p.150の「製版・目打組合せ型式」を参照ください。

⓰定常変種の表記方法
本カタログでは、定常変種にA-pos.1やC-pos.10のように、実用版の中でのシート位置（仮）を示す大文字アルファベットを付け、その後に切手のシート内位置を示す記号を表記し

＊＊：extremely fine condition (unused)
◉：fine condition (used)
The values may also use the following symbols.
u：unknown
x：non-existent
−：not determined

❾**Catalogue Values of FDC**
The values of the first day covers are shown in yen.

❿**Printing**
The applicable printing methods are shown here.

⓫**Perforation**
The applicable perforation types are listed here.

⓬**Printed Area Size**
The printed area refers to the limits of the stamp design and face value indication, not the size of the stamp. It is shown in horizontal x vertical order, expressed in millimeters.

⓭**Number Issued**
In the case of souvenir sheets and booklet panes, the number of copies is indicated.

⓮**Pane Format**
The number is the number of stamps in a pane, and two numbers in (　) are the number of stamps in row and column of pane. (number in row x number in column).

⓯**Plate and Perforation Configuration**
This catalogue shows combinations of plate

年賀切手の例 New Year's Greeting Stamps

⑤ N14 / N14A　1959年（昭和34）用
For 1959 (Showa 34)

発行日：N14=1958（昭和33）.12.20、N14A=1959（昭和34）.1.20
Date of Issue：N14=1958 (Showa 33).12.20、N14A=1959 (Showa 34).1.20

N14 鯛えびす
（高松の郷土玩具）
Papier Mache "Tai-ebishu" (Folk Toy of Takamatsu)

N14A お年玉小型シート
"New Year's Gift" Souvenir Sheet (lottery prize)

1) 各刷色では、1版と2版に共通の定常変種は確認できていない。
1) No common constant flaw of each color between plates 1 and 2 is known.

● スクリーン角度 Screen Angles　　1版 Plate 1　赤60度 red 60 degrees　　2版 plate 2　赤65度 red 65 degrees

 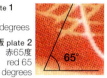

③ ●風景印 Picture Postmark

高松 Takamatsu

⑤ N14 ● 定常変種 Constant Flaws

P1-A-pos.1　P1-A-pos.12　P1-B-pos.11　P1-B-pos.19

⑯

  （追加画像）

P2-A-pos.1　P2-A-pos.19　P2-B-pos.14　P2-B-pos.16

⑤	⑥	⑦		⑧	⑨
			**	●	FDC
N14	5 yen	多色 multicolored		60	40	800
N14A	20 yen (5 yen × 4)			1,300	1,500	3,500

切手データ Basic Information	
N14	
⑬発行枚数：1,500万枚	number issued : 15 million
⑮製版・目打組合せ：タイプ1B-4	plate and perforation configuration : type 1B-4
N14A	
⑬発行枚数：1,044.2万枚	number issued : 10,442,000

ています。また、実用版が複数ある場合は、P1やP2のように実用版番号を付記して、P1-A-pos.1のように表記します。

4.日本語の表記
学術研究の観点から、切手、地名などの表記は当時の呼称で示します。例えば、国際的に承認されていない国名（満州国など）、現在では一般に採用されない地理的名称（支那など）を使用することがあります。

常用漢字を使用することを基本としていますが、専門表記については常用漢字以外の漢字を使用することがあります。

configuration, perforation types, and pane configurations of perforation for each stamp. Details can be found in "Configurations of Plate and Perforation" on p. 150.

⑯ Notation of Constant Flaws
In this catalogue, constant flaws are identified with capital letters indicating the pane position (tentative) in the plate followed by the stamp position in the pane, such as A-pos.1 or C-pos.10. In addition, if there are multiple plates, the plate number is added, such as P1 or P2, and indicated as P1-A-pos.1.

カタログ番号の分野別記号 Classification of Stamps and Postal Stationery （日本郵趣協会監修のカタログ共通）

分野 Category		記号 Prefix	分野 Category		記号 Prefix
■切手 Stamps			■ステーショナリー Postal Stationery		
記念・特殊	Commemorative and Special Stamps	C	記念・特殊はがき	Commemorative and Special Postal Cards	CC
記念切手帳	Commemorative Booklets	CB	年賀はがき	New Year's Greeting Postal Cards	NC
記念切手帳ペーン	Commemorative Booklet Panes	CP	季節見舞はがき	Seasonal Greeting Postal Cards	SG
公園	National Park Stamps	P			
年賀	New Year Greeting Stamps	N			

記念・特殊切手 Commemorative and Special Stamps 1952-1966

■記念・特殊切手の特徴

本カタログで採録している記念・特殊切手(ここでは公園切手、年賀切手を含む)は、その発行時期はが普通切手の動植物国宝図案切手の発行時期と重なるため、印刷方式、実用版上のシート配置、目打型式、用紙、裏糊、グラビア版のスクリーン線数・方向など製造面の諸要素は、普通切手と共通している。

本カタログでは、これらの製造面のバラエティを極力網羅している。特にグラビア印刷の切手では、目打の穿孔方式、定常変種、スクリーンの要素から実用版上のシート配置までを検証して、郵便局から発売された切手シートを研究分類している。

表1は、採録した記念・特殊切手の印刷に関わる基本的事項をまとめたものである。

■Features of Commemorative and Special Stamps

Since the printing period of all the stamps in this catalogue is almost same as that of the 1950-1965 Fauna, Flora & National Treasure Series definitive stamps, the details of the manufacturing process, such as pane configurations of plate, perforation types, papers, gums, photogravure screen numbers and angles are common between them.

This catalogue covers as many of these manufacturing details as possible. The photogravure printed stamps, in particular, are intensively investigated, with explanations of perforation types, constant flaws, and photogravure screens.

表1 印刷に関する基本的事項 Table 1 Fundamental Classification of Manufacturing Process

版式 plate type	印刷機 printing press	実用版のシート配置 pane configurations of plates	目打型式 perforations	用紙 papers
グラビア版 photogravure	板グラビア機 flat plate photogravure press	頭合せ、順並び(通常2×2面田型) head-to-head or normal (normally 2×2 panes)	普通櫛型 normal comb perf.	枚葉紙 sheet paper
	多色刷輪転機 multi-color rotary photogravure press	縦並び、横並び(通常4面) vertical or horizontal (normally 4 panes)	連続櫛型(一部全型) continuous comb perf. (some with harrow perf.)	連続紙 web paper
凹版 intaglio	速刷機 high-speed intaglio press	頭合せ、順並び(通常2×2面田型) head-to-head or normal (normally 2×2 panes)	普通櫛型 normal comb perf.	枚葉紙 sheet paper
	小型凹版輪転機 small rotary intaglio press			

1. 印刷版式と印刷機

記念・特殊切手の印刷版式には、グラビア印刷と凹版印刷がある。グラビア印刷では当初、実用版を平らな銅板上に製版して印刷の際に輪転機に巻き付ける板グラビア機が使用されていた。1954年(昭和29)にドイツから実用版を円筒状の版胴に直接製版する多色刷輪転機が輸入されると、最初に普通切手の印刷に使用が開始され、記念特殊切手では、同年12月発行の年賀切手「加賀八幡起き上がり」(N10)の2色刷り印刷に初めて使用された。

以降、グラビア印刷の記念・特殊切手には、単色刷りであっても多色刷輪転機が使用された。その後、印刷ユニットが追加されて4色まで連続印刷できるようになり、1955年(昭和30)年5月発行の「第15回国際商業会議所総会記念」(C249)が最初の4色刷り印刷の切手となった。

凹版印刷では、従来から正方形の印刷装置に最大4面の実用版を取り付けることが可能な凹版速刷機が使用されていた。1957年(昭32)以降、同一版面に3色のインクを

1. Printing Methods and Presses

The printing methods for the commemorative and special stamps in this catalogue are photogravure printing and intaglio printing. In the beginning, photogravure printing used the flat plate process, in which the design of the stamp was transferred to a flat copper plate which was curved around a rotary drum for printing. In 1954 (Showa 29) a multi-color rotary press, on which the design was transferred directly to a cylindrical plate, was imported from Germany and was used at first to print definitive stamps. For commemorative and special stamps, it was first used for the two-color printing of the New Year's Greeting Stamp "Kaga Hachiman Okigari Doll" (N10) issued in December of the same year.

The multi-color rotary press was used for photogravure printed commemorative and special stamps thereafter, even for single-color printing.

N10 加賀八幡起き上がり
Kaga Hachiman
Okiagari Doll

C249 第15回国際商業会議所総会記念
15th Congress of International
Chamber of Commerce

詰めて印刷できる「ザンメル凹版印刷」が可能な小型凹版輪転機が開発されると、凹版速刷機と使い分けて使用された。

2. 実用版上のシート配置

グラビア版でも凹版でも、いくつかの例外を除いて1枚の実用版にはシート4面が田型に配置された。当初は上段の2シートが天地逆となる「頭合せ」(図1)で製版されていたが、1958年(昭和33)になると上段・下段とも同じ向きになる「順並び」(図2)で製版されるようになった。同年8月発行の「佐渡弥彦国定公園」(P200-201)以降は、上段・下段とも同じ向きになる「順並び」で製版されたことが目打穿孔方式から判明している。ただし、同年3月発行の「関門トンネル開通記念」(C272)から6月発行の「ブラジル移住50年記念」(C279)までは、目打穿孔方式でも判別できないため一応「頭合せ」としているが「順並び」の可能性もある。なお、「ビードロを吹く娘」(C252)以降の切手趣味週間だけは例外で、すべて「順並び」である。

多色刷輪転機によるグラビア版では、連続櫛型目打で穿孔できるようになったため、シート配置は1961年(昭和36)年4月発行の花シリーズ「ヤマザクラ」(C330)以降、縦型切手は4面横並び、横型切手は4面縦並びで製版され

Later, a printing unit was added to make it possible to print up to four colors continuously. The first four-color stamp was the "15th Congress of International Chamber of Commerce (ICC)" (C249), issued in May 1955.

For intaglio printing, a high-speed Intaglio press capable of mounting up to four plates on a square mount had been in use. After 1957 (Showa 32), a small rotary intaglio press capable of "sammeldruck intaglio printing" was developed, in which the plate is filled with three colors of ink and attached to the plate cylinder of a rotary press in the same way as a photogravure plate. Either the high-speed intaglio press or the small rotary intaglio press was used for printing, according to individual stamp characteristics.

2. Pane configurations on the Plate

All pane configurations on the plates were 2× 2, with some exceptions, for both photogravure and intaglio printing. In the beginning, a head-to-head configuration (Fig. 1) was adopted; after 1958 (Showa 33), the normal configuration (Fig. 2) was adopted. It has been determined that the normal configuration was first adopted on the "Sado-Yahiko National Park" stamps (P200-P201), issued in August 1958. Although C272 to C279, which were issued from March to June 1958, are listed as head-to-head configuration due to uncertainty of perforation type, it is possible that the normal configuration is applicable to these stamps. All the Philatelic Week stamps, beginning with "Woman with a Glass Noisemaker (Poppen)" (1955), are of the normal configuration.

Since the continuous comb perforator was

図1 頭合せ Fig.1 head-to-head configuration

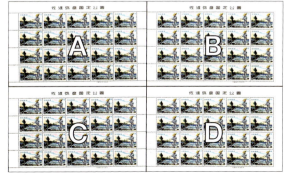
図2 順並び Fig.2 normal configuration

記念・特殊切手

るようになった（一部の切手では3面、6面並び）。切手によっては2×2面順並びも採用されていて使い分けの基準は明確ではないが、同一切手の印刷に両方式が混用された実例はない。

3. 目打型式

目打型式は普通櫛型、連続櫛型、全型の3種類あり、ピッチはいずれも13½が主であるが、一部の切手では13、13×13½、12½等が使われている。基本的には、4面田型の実用版で印刷された切手は普通櫛型で、縦並びまたは横並びの実用版で多色刷輪転機により印刷された切手は連続櫛型である。実用版と目打穿孔方式は密接に関係しており、その組合せをパターンで分類して整理したのが巻末資料「製版・目打組合せ型式」（p.150～154）である。

印刷後に別工程で目打穿孔機にかける場合は、1面ずつ裁断してから穿孔、縦または横に2面続きに裁断してから穿孔、4面田型のまま穿孔の3つの方式があり、耳紙への目打の抜け方によりさらに細分化される。

連続目打は、多色刷輪転印刷機に付属した目打穿孔機によって印刷、目打穿孔の工程を連続して行うもので、この方式が実用化されたことによって実用版構成は縦並びまたは横並びに変更された。なお、一部の切手と年賀切手小型シートは、全型目打による穿孔である。

目打の継ぎ目は、連続櫛型目打の二連目打の場合、縦型切手では縦目打との交差部に、横型切手では横目打との交差部にそれぞれ1列おきに現れる。櫛歯の向きは、用紙の流れの上流に向いているもの（右向き又は下向き）と下流に向いているもの（左向き又は上向き）がある。その特徴は、右向き（下向き）では縦（横）の目打列の左側

図3　連続目打の櫛歯向きの混用例（N19、P1-A-pos.19／20）
Fig.3 Example of both comb teeth directions usage on same stamp（N19）

右向き（左側の間隔狭い）right facing (left space is narrow)

左向き（右側の間隔狭い）left facing (right space is narrow)

adopted for the multi-color rotary photogravure press, the pane configuration was changed to 4-pane horizontal for vertical stamps or 4-pane vertical for horizontal stamps starting with the "Mountain Cherry, Flower Series" (C330) issued in April 1961 (Showa 36). (a 3-pane or 6-pane configuration was adopted for some stamps). Even during this period, the 2×2 configuration was adopted for some stamps but both the 2×2 configuration and 4-pane configuration were not used for the same stamp.

3. Perforations

Three types of perforators were in use during the period covered by this catalogue: comb, continuous comb, and harrow. The gauge of all three perforator types was 13½. The type of perforator was basically determined by the pane configuration of the plate: comb perforator for 2×2 configuration, continuous comb perforator for vertical or horizontal configuration printed on the multi-color rotary press. The plate and perforation configurations were closely related, and their combinations are summarized in the Appendix "Plate and Perforation Configurations" (pp. 150-154).

When the sheet was perforated in a separate process after printing, there were three methods: cutting to single pane and then perforating, cutting to vertical or horizontal pair of two panes and then perforating, and perforating an entire 2×2 sheet; these methods are further sub-classified according to the perforation configuration in the selvage.

For continuous perforation, printing and perforation were executed together on the multi-color rotary press with attached perforator. Following implementation of this method, the pane configuration of plates was changed to vertical column or horizontal row. However, some stamps and the souvenir sheets of New Year's Greeting stamps were perforated with the harrow perforator.

With continuous double comb perforation, the joints of the perforations appear at every other column of horizontal perforations for vertical stamps and at every other row of vertical perforations for horizontal stamps. The teeth of the comb are oriented upstream (rightward or downward) or downstream (leftward or

（上側）に出て、左向き（上向き）では縦（横）の目打列の右側（下側）に出てくる。同一の切手で櫛歯の向きが混用されている場合もある（図3）。

目打の向きが同一切手で混用された例は、田型4面で櫛型穿孔された凹版印刷の「東京1964オリンピック競技大会（寄附金付）」第5次、第6次（C359-362、C363-366）でも確認されている。

4. 用紙

終戦直後は「灰白紙」と呼ばれる粗末な紙であったが、書状料金10円時期にはカレンダー（つや出し機）にかけられた滑面のある「白紙」が使用されるようになった。多色刷輪転機が輸入されるとグラビア印刷との適性が研究され、同機による印刷第1号となった1954年（昭和29）12月発行の年賀切手「加賀八幡起き上がり」（N10）から印刷局王子工場製の「75SC」、1963年（昭和38）発行の鳥シリーズ以降は更に改良された「改正75SC」という用紙が使用された。

板グラビア機、凹版速刷機及び小型凹版輪転機では、「枚葉紙」といわれるカット紙が使用された。印刷された切手には用紙の漉き目が縦になる「縦紙」、漉き目が横になる「横紙」があるが、切手によっては用紙の入れ方を逆にしたことが原因で混在する場合がある。グラビア版の多色刷輪転機は連続紙といわれるロール紙が使用されているので、紙の進行方向と同じ向きで製版されたものは「縦紙」、進行方向に対して横向きに製版されたものは「横紙」となる。

5. 裏糊

1951年（昭和26）頃はアラビアゴム糊（A糊）、1952年（昭和27）からはA糊にデキストリンを加えたアラビアゴム・デキストリン糊（AD糊）が使用されていたが、日本の夏の高温多湿の気候では重ねたシートが接着し易いため、シートの間にわざわざグラシン紙を挟んで切手同士の接着を防止していた。その後、接着しにくいポリビニール・アルコール糊（PVA糊）が開発されて、1958年（昭和33）頃から、普通切手に先駆けて使用されるようになった。これにより、グラシン紙の挟み込みがなくなった。

しかし、PVA糊はAD糊よりも弾性が低いため、この頃に発行された切手には目打の抜けが悪いものが多い。アルコールの調合を変えて改良が重ねられ、1963年（昭和38）後期からは艶があり乾燥度を高めたPVA糊が使用された。

6. グラビア版のスクリーン線数、角度

板グラビア機では、スクリーン200線、45度が使用された。多色刷輪転機の実用版ではスクリーン260線であるが、モ

upward) of the printing flow. The teeth of the comb are on the left (top) side of the vertical (horizontal) perforations when oriented to the right (downward), and on the right (bottom) side of the vertical (horizontal) perforations when oriented to the left (upward). There are also cases where the two teeth directions are used on the same stamp. (Fig. 3).

Examples with both directions of the comb teeth on the same stamp are also confirmed for C359-C362 and C363-C366, which were perforated with the comb perforator on 2×2 panes.

4. Papers

A low-quality paper called "coarse white paper" was used for a short time after the end of the Second World War, followed by a smooth-surfaced (calendered) "white paper" which came into use during the 10 yen letter rate period. After introduction of the imported multi-color rotary press, research was conducted into a better combination between paper and photogravure printing; for the first printing with this press, "Kaga Hachiman Okiagari Doll, New Year's Greeting Stamp" (N10, December 1954), paper "75SC" made by the Printing Bureau's Oji factory was used. Beginning with the "Bird Series" issued in 1963 (Showa 38), a further improved "revised 75SC" paper was used.

Sheet papers were used for the flat plate photogravure press, high-speed intaglio press and small intaglio rotary press. There were two types of sheet papers: long grain and short grain. Both types of paper were used for some stamps. Since web paper was used for the multi-color rotary photogravure press, the type of paper was specific to the direction of printing.

5. Gums

Arabic-based gum (A gum) was in use around 1951 (Showa 26), followed by Arabic-Dextrin gum (A-D gum), in which Dextrin was added to gum arabic, from 1952 (Showa 27). At the time, glassine papers were inserted between the panes to prevent stamps from sticking together due to the hot and humid summer climate of Japan. A non-sticky Polyvinyl Alcohol gum (PVA gum) was developed and began to be used around 1958 (Showa 33), ahead of its use on definitive stamps.

図4 スクリーン線数（C404）Fig.4 Screen Line Number for C404

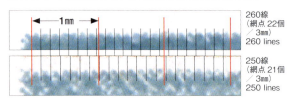

260線（網点22個／3mm）260 lines
250線（網点21個／3mm）250 lines

アレ防止の理由で各刷色ごとに異なる方向に設定された。また同じ刷色にも角度変化があるため、組合せのバラエティが存在する。それらは各切手の項で注記したほか、年賀切手については p.143 に示した。

1964年（昭和39）7月発行のお祭りシリーズ「祇園祭」（C404）では青色250線のスクリーンが260線と併用され、以後のグラビア切手はすべて250線に統一された（**図4**）。また、1965年（昭和40）4月発行の切手趣味週間「序の舞」（C425）以降の切手は、スクリーン角度はすべて45度に統一された。

スクリーン角度は測定時の誤差が一定程度あり、現在も研究途上にある。しかし、同一切手でのスクリーン角度の違いは実用版が異なることを示すものであり、版分類には欠かせない要素である。

なお、スクリーン角度の数値は参考文献「郵趣モノグラフ31 製造面から見た書状10円期の記念特殊切手」の記載と異なる場合があるが、同書の著者と検討・修正を行って本カタログでは最新の数値を掲載している。

7. 実用版上のシート配置の呼称

窓口シートの分類は、定常変種とスクリーン角度に基づいている。ただし、それらの実用版上のシート配置は、印刷方式と目打穿孔方式を加味して、あくまでも推定による仮の位置として定めたものであり、実際の配置位置を特定することは不可能である。ただし、鳥シリーズ「ルリカケス」（C390）

図5 印刷用トンボの位置（C390）Fig.5 Marginal Guide Marks for C390

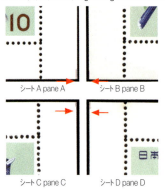

シート A pane A　シート B pane B
シート C pane C　シート D pane D

With the introduction of PVA gum, the insertion of glassine papers between panes was abolished.
On the other hand, many stamps of this period were characterized by poor perforations because PVA gum was less elastic than A-D gum. The quality of PVA gum was improved by changing the proportion of alcohol, and high quality PVA gum began to be used starting in late 1963 (Showa 38).

6. Photogravure screen numbers and angles

A 45-degree, 200-line photogravure screen was used for the flat-plate photogravure press plates. For the multi-color rotary press plates, the screen number was 260 and the screen angle of each color was different to prevent a moiré effect. Since the screen angle varied even for same color, there were variations in the combination of screen angles; this is explained in the individual stamp sections of this catalogue. The screen angle variations for the New 's Greeting Stamps are shown on p. 143.
For "Gion Festival, Festival Series" (C404) issued in July 1964 (Showa 39), two screen line numbers of 250 and 260 were used for the blue plate (Fig. 4). Thereafter, the screen number was unified to 250. The screen angles of all colors were unified to 45 degrees for "Philatelic Week, Japanese Dance: Jo-no-mai by Uemura Shoen," (C425) issued in April 1965 and thereafter.
There is always some error in measurement of screen angles, and research on screen angles is still ongoing. As screen angle differences indicate plate differences, screen angle measurement is an essential element for plate classification.
Some screen angles are different from those listed in the reference book "Philatelic Monograph 31: Commemorative Stamps of the Letter 10 yen period from the Viewpoint of Manufacturing". In that case, the screen angles in this catalogue are the latest values.

7. Pane Position in Plate

The classification of panes in this catalogue is based on distinctions of constant flaws and screen angles. On the other hand, the positions of panes in the plate are deduced from printing and perforation methods. Therefore, pane positions in this catalogue are tentatively determined and the true positions are unknown. As the only exception,

と「ライチョウ」(C391) は、耳紙の四隅に残された裁断用のトンボの位置からシート配置が特定できる唯一の例である（図5）。

シート面の呼称は、田型4面掛けの場合は頭合せか順合せかに関わらず、上段は左からA、B、下段は左からC、Dとしている。多色刷輪転機で横または縦に並べて配置された連続目打の場合は、継目が縦型切手では縦目打との交差部に、横型切手では横目打との交差部に現れる。二連目打では、それぞれ1列置きに現れるので、最左列または最上段の目打交差部に継目が現れるものをA、以降B、C、Dとしている（3面掛けの場合はA～C、6面掛けの場合はA～F）。なお一部の切手では、継目が1列分ずれたものが存在する（図6）。

図6 二連目打の継目位置バラエティ (C421)
Fig.6 Perforation Joint Variety of Double Comb Perforation for C421

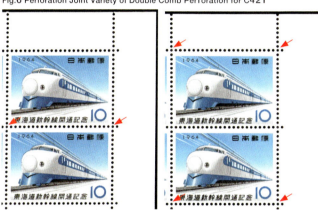

1版、シートA、ポジション5/10
Plate 1, Pane A, pos.5/10

定常変種は同一シートに複数確認されているが、紙面の都合によりシート分類において鍵となる最も明瞭なものを2点程度掲載している。

また、凹版切手及び凹版グラビア切手では、凹版の実用版の製造工程では定常変種が発生しにくく、現時点では「小河内ダム竣工記念」(C270)と「第15回国際航空運送協会総会記念」(C300)の2種の確認にとどまっている。

8. ご当地局の定義と類型

使用例においては「ご当地局」の概念を取り上げ、巻末資料「ご当地局の定義と類型」(p.155～160)において詳述した。評価という点では未だ定着していないが、評価上の観点を示している。

the true pane positions on the plate are known for "Purple Jay" (C390) and "Rock Ptarmigan" (C391) of the Bird Series. The marginal guide marks at the center of the four panes in a plate remained in some panes, making it possible to determine the pane position (Fig. 5).

The positions of pane for 2×2 configuration are called "A", "B" in the upper row from left to right, and "C", "D" in the lower row from left to right, regardless of whether they are head-to-head configuration or normal configuration. In case of double comb perforations in vertical column configuration or horizontal row configuration on the multi-color rotary press, a perforation joint appears in every other column for vertical stamps or every other row for horizontal stamps. The pane in which the joint appears at the leftmost perforation for vertical stamp or at the top perforation for horizontal stamp is thus called "A", followed by "B", "C", "D". (In case of 3-pane and 6-pane configurations, "A" - "C", "A" - "F", respectively) In addition, some panes have a joint that is offset by one row or column (Fig. 6).

Although multiple constant flaws are known to exist in the same pane, only a couple of the clearest are listed in this catalogue due to space limitations.

Generally, constant flaws are scarce for intaglio printed stamps because the plates of intaglio printing are made by the transfer method. So far, the only two confirmed constant flaws are in "Completion of Ogouchi Dam" (C270) and "15th General Meeting of International Air Transport Association" (C300).

8. Definition and Types of Related Area Post Office

In this catalogue, the concept of "Related Area Post Office" is taken up as one of the categories of usage. A detailed discussion of "Related Area Post Office" is presented in Appendix (pp. 155 - 160). Although this has not yet been established as a criterion for evaluation, the appendix provides a perspective for that purpose.

(1) 額面「銭位」時期 "Sen" Value Era 1952

C225-226 万国郵便連合加入75年記念
75th Anniversary of Admission to Universal Postal Union (UPU)

発行日：1952（昭和27）.2.19
Date of Issue：1952 (Showa 27) .2.19

C225 南十字星
Southern Cross

C226 北斗七星
Big Dipper

●特印
Special Date Stamp

C227-228 東京 Tokyo

	**	●	FDC
C225 5 yen 暗い青味紫 purple	1,400	300	
C226 10 yen 暗い青緑 dark green	3,700	800	
C225-226 (2種 set of 2 values)	5,100	1,100	5,500

切手データ Basic Information

版式：板グラビア版	printing plate: flat photogravure plate
目打：櫛型13½	perforation: comb perf. 13½
印面寸法：22.5×33mm	printing area size: 22.5×33mm
発行枚数：各300万枚	number issued: 3 million each
シート構成：20枚（横4×縦5）	pane format: 20 subjects (4×5)
銘版、銘版位置：⑫印刷庁（新庁）19番	imprint and position: type ⑫ pos.19
製版・目打組合せ：タイプ1A-3	plate and perforation configuration: type 1A-3

C225 ●定常変種 Constant Flaws

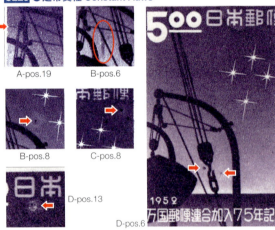

A-pos.19　B-pos.6　B-pos.8　C-pos.8　D-pos.13

位置 position	特徴 feature	**	●
A-pos.19	滑車の左に紫点群 purple dots to left of pulley	3,000	800
B-pos.6	ロープに帯状の線 striped line on rope		

右上につづく Continued to up right ↗

B-pos.8	南十字星の下方に大紫点 large purple dot below Southern Cross		
C-pos.8	"便"の下に紫点 purple dot below "便"		
D-pos.6	フックの左右に点 dots to left and right of hook		
D-pos.13	"日"の下に白ぼやけ white blur below "日"		

C226 ●定常変種 Constant Flaws

A-pos.3　A-pos.11　B-pos.1　B-pos.3

位置 position	特徴 feature	**	●
A-pos.3	"日"に緑点 green dot on "日"	10,000	2,000
A-pos.11	"日"の左下方に濃い緑点とインクの流れ dark green dot and ink flow to lower left of "日"		
B-pos.1	"念"の右上に大きな欠損 large defect to upper right of "念"		
B-pos.3	"日"の左上に大きな濃い緑点 large dark green dot to upper left of "日"		

(2) 額面「円位」時期、ローマ字なし "Yen" Value Era, without NIPPON 1952-1965

C227-228 日本赤十字社創立75年記念
75th Anniversary of Japanese Red Cross

発行日：1952（昭和27）.5.1
Date of Issue：1952 (Showa 27) .5.1

（左）C227
赤十字とヤマユリ
(left) Red Cross and Lilies

（右）C228 看護師
(right) Red Cross Nurse

●特印
Special Date Stamp

東京 Tokyo

	**	●	FDC
C227 5 yen 赤 rose red and dark red	1,200	350	
C228 10 yen 暗い青味緑・赤 dark green and red	2,800	750	
C227-228 (2種 set of 2 values)	4,000	1,100	6,000

記念・特殊切手（額面「円位」時期、ローマ字なし）

切手データ Basic Information	
版式：板グラビア版	printing plate: flat photogravure plate
目打：櫛型13½	perforation: comb perf. 13½
印面寸法：22.5×33mm	printing area size: 22.5×33mm
発行枚数：各300万枚	number issued: 3 million each
シート構成：20枚（横4×縦5）	pane format: 20 subjects (4x5)
銘版・銘版位置：⑫印刷庁（新庁）19番	imprint and position: type ⑫ pos.19
製版・目打組合せ：タイプ1A-3	plate and perforation configuration: type 1A-3

C227 ●定常変種 Constant Flaws

 A-pos.5   A-pos.13   B-pos.4

位置 position	特徴 feature	**	●
A-pos.5	赤十字の左に赤いシミと大きな白抜け red stain and large white spot to left of Red Cross	3,000	1,000
A-pos.13	ユリに花粉 pollen to left of lilies		
B-pos.4	"5"の左下に赤点 red dot to lower left of "5"		

C228 ●定常変種 Constant Flaws

 A-pos.7
 A-pos.14
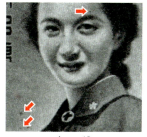 A-pos.18
 B-pos.1

位置 position	特徴 feature	**	●
A-pos.7	頭の左に白抜け white spot to left of head	6,000	2,000
A-pos.14	頬に大きい濃い点 large dark dot on cheek		
A-pos.18	眉毛の上に緑点、肩の左に緑点2つ green dot above eyebrow, two green dots to left of shoulder		
B-pos.1	頭の右に緑点2つ、肩の右に緑点 two green dots to right of head, green dot to right of shoulder		

C229 東京大学創立75年記念 / 75th Anniversary of Tokyo University

発行日：1952（昭和27）.10.1   Date of Issue : 1952 (Showa 27).10.1

C229 安田講堂
Yasuda Hall, Tokyo University

●特印 Special Date Stamp

東京 Tokyo

	**	●	FDC
C229 10 yen 暗い灰緑 dull green	4,200	600	3,800

切手データ Basic Information	
版式：凹版速刷版	printing plate: high-speed intaglio printing
目打：櫛型13½	perforation: comb perf. 13½
印面寸法：22.5×27mm	printing area size: 22.5×27mm
発行枚数：300万枚	number issued: 3 million
シート構成：20枚（横4×縦5）	pane format: 20 subjects (4x5)
銘版：⑬大蔵銘（C230以降同様、記載省略）	imprint: type ⑬ (Omitted as it is the same for C230 and later)
製版・目打組合せ：タイプ1A-3	plate and perforation configuration: type 1A-3

C230-231 第7回国体記念 / 7th National Athletic Meet

発行日：1952（昭和27）.10.18   Date of Issue : 1952 (Showa 27).10.18

C230 山岳競技
Mountain Climbing

C231 レスリング
Wrestling

●特印 Special Date Stamp
C230-231
福島 Fukushima

	**	●	FDC
C230 5 yen 紫青 blue	1,500	400	
C231 5 yen こい茶 dark brown	1,500	400	
C230-231 (2種連刷 pair)	4,000	2,000	4,800

切手データ Basic Information	
版式：板グラビア版	printing plate: flat photogravure plate
目打：櫛型13½	perforation: comb perf. 13½
印面寸法：27×22.5mm	printing area size: 27×22.5mm
発行枚数：各200万枚	number issued: 2 million each
シート構成：20枚（横5×縦4）	pane format: 20 subjects (5x4)
製版・目打組合せ：タイプ1A-2	plate and perforation configuration: type 1A-2

●目打の抜け方 Position of Perforations Through Selvages

正常 normal

逆抜け irregular

縦2面頭合わせで目打穿孔を行う際に、シート天地を逆にして目打機に装着したため、目打の逆抜けが存在する。左図はいずれも定常変種A-pos.1（肩の左に定常変種）があるのでシートAであり、上図が正常で下図が逆抜けになる。 Some sheets were set upside down when they were perforated, which resulted in reverse perforation. Since constant flaw A-pos.1 exists in both figures at left, these two are sheet A. The upper figure is normal perforation, whereas the lower figure is reverse perforation.

記念・特殊切手（額面「円位」時期、ローマ字なし）

C230-231 ●定常変種 Constant Flaws

A-pos.1

A-pos.20

B-pos.4

B-pos.15

B-pos.7

C-pos.2 / C-pos.7

位置 position	特徴 feature	**	●
A-pos.1	肩の左に白抜けとインクの流れ、腕の左に青点 white spot and ink flow to left of shoulder, blue dot to left of arm	4,000	1,000
A-pos.20	背中に茶点3個 three brown dots on back		
B-pos.4	"体"の下に太い茶線 thick brown line below "体"		
B-pos.7	腕の右に青い線と点、"5"の左方と下に青点 blue line and dot to right of arm, blue dots to left and below "5"		
B-pos.15	腕の右に斜線と青いシミ diagonal line and blue stain to right of arm		
C-pos.2	背中と脚に茶点 brown spots on back and leg		
C-pos.7	脚に濃い青点、ピッケルの左に青点 dark blue dot on leg, blue dot to left of ice axe		

C232-235 明仁立太子礼記念
Ceremony for Proclamation of Crown Prince Akihito

発行日：C232-234 1952（昭和27).11.10
C235 1952（昭和27).12.23

Date of Issue：C232-234 1952 (Showa 27).11.10
C235 1952 (Showa 27).12.23

C232、C233
麒麟と菊
Qilin and Chrysanthemum

C234
皇太子旗
Flag of Crown Prince

●特印 Special Date Stamp

東京 Tokyo

```
                              **      ●     FDC
C232  5 yen だいだい・暗い赤紫 red-orange and purple
                              500    200
C233 10 yen 黄味だいだい・こい緑 red-orange and dark green
                              600    200
C234 24 yen 暗い青 deep blue   3,200  1,600
C232-234 (3種 set of 3 values) 4,300  2,100  4,000
C235 小型シート (売価50円) souvenir sheet (selling price 50 yen)
                            18,000 17,000 100,000
C235a  5 yen 目打・糊なし imperforate, ungummed
                             3,000  4,000
C235b 10 yen 目打・糊なし imperforate, ungummed
                             3,500  4,000
C235c 24 yen 目打・糊なし imperforate, ungummed
                             7,000  9,000
```

1) 5円と10円のだいだい色のグラビア印刷部分は定常変種が全く同じであることから、同一の実用版が使用されたと推定できる。
1) Since the constant flaws in the red-orange printed area are identical on the 5 yen and 10 yen stamps, it is thought that the same photogravure plate was used for both.

C235 小型シート Souvenir Sheet

半透明袋 Translucent Paper Envelope

C232-234 切手データ Basic Information	
版式：C232、C233=板グラビア・凹版速刷版、C234=凹版速刷版	printing plate：C232,C233=flat photogravure plate and high-speed intaglio printing, C234=high-speed intaglio printing
目打：櫛型13½	perforation：comb perf. 13½
印面寸法：C232、C233= 22.5×27mm、C234=27×22.5mm	printing area size：C232, C233=22.5×27mm, C234=27×22.5mm
発行枚数：C232、C233=各500万枚、C234=100万枚	number issued：C232, C233=5 million each, C234=1 million
シート構成：C232、C233=各20枚（横4×縦5）、C234=20枚（横5×縦4）	pane format：C232, C233=20 subjects of (4x5) each, C234=20 subjects (5x4)
製版・目打組合せ：C232、C233=タイプ1A-3、C234=タイプ1A-2	plate and perforation configuration：C232, C233=type 1A-3, C234=type 1A-2

C235 切手データ Basic Information	
版式：板グラビア・凹版	printing plate：flat photogravure plate and intaglio printing
目打・糊なし	perforation：imperforate, ungummed
シート寸法：130×130mm	sheet size：130×130mm
発行枚数：15万枚	number issued：150,000

記念・特殊切手（額面「円位」時期、ローマ字なし）

C232-233 ●定常変種 Constant Flaws

A-pos.5　　B-pos.4　　C-pos.2　　D-pos.13

位置 position	特徴 feature	**	●
A-pos.5	第3コーナーにだいだい点3個 three orange dots in SE corner	1,200	500
B-pos.4	第2コーナーに大きなだいだい点 large orange dot in NE corner		
C-pos.2	下マージンにシミ状の点 smudge in bottom margin		
D-pos.13	第4コーナーに濃い帯 dark band in SW corner		

●実用版構成 Plate Configuration

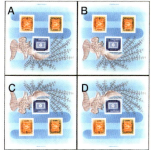

4面掛け（2×2面頭合せ）
4 panes (2×2-pane head-to-head configuration)

1) 小型シートは4面掛け（2×2面頭合せ）で印刷されたが、シート地の定常変種により2つのタイプに区分される。しかし、シート地については4面掛けの実用版が2版存在したのか、4面掛けを並べて8面掛けで印刷した後に4面掛けに半裁したのかは定かではない。

1) The souvenir sheet was printed in 2×2-pane configuration, which is classified into two types according to the constant flaws in the background. However, it has not been determined whether there were a) two individual 4-pane plates or b) one 8-pane plate consisting of two joined 4-pane plates, the stamps being divided into two sheets of four panes each after printing.

C235 ●定常変種 Constant Flaws

印面 printing area		シート地 background	

A

5 yen 下マージンにだいだい点 / orange dot in bottom margin　　10 yen 下マージンに大きいだいだい点 / large orange dot in bottom margin　　type 1 濃い青地に白抜けと小さな青点 / white spot and small blue dot in dark blue background　　type 2 濃い青地と薄い青地に白抜け / white spots in dark blue and light blue backgrounds

B

5 yen 第2コーナーにだいだい点 2個 / two orange dots in NE corner　　type 1 シミ状のリタッチと白抜きのリタッチ / stain-like retouch and white-spot retouch　　type 2 白抜けのリタッチ 2ヵ所 / two white-spot retouches

C

5 yen 左側にだいだい点 / orange dot on left side　　type 1 青地に大きなリタッチ 2ヵ所 / two large retouches in blue background　　type 2 青地に縦筋の大きなリタッチ 2ヵ所 / two large retouches with vertical lines in blue background

D

5 yen 右側にだいだい点 2ヵ所 / two orange dots on right side　　type 1 濃い青地に大きな白抜け / large white spot in dark blue background　　type 2 薄い青地に濃い青点 / dark blue spot in light blue background

C236 電灯75年記念 / 75th Anniversary of Electric Lamp
発行日：1953（昭和28）.3.25　Date of Issue : 1953 (Showa 28).3.25

C236 最初のアーク灯 / First Arc Lamp

●特印 Special Date Stamp

東京 Tokyo

C236 ●定常変種 Constant Flaws

A-pos.3　　A-pos.16　　B-pos.1　　B-pos.14

	**	●	FDC
C236 10 yen こい茶 brown	1,500	500	3,200

位置 position	特徴 feature	**	●
A-pos.3	アーク灯の左に斜線と白抜け diagonal line and white spot to left of lamp	4,000	1,500
A-pos.16	アーク灯の右下に大きな濃い点 large dark dot to lower right of lamp		
B-pos.1	アーク灯の右に濃い点と薄い点2個 dark dot and two thin dots to right of lamp		
B-pos.14	アーク灯に大きな濃い点 large dark dot in lamp		

切手データ Basic Information

版式：板グラビア版	printing plate : flat photogravure plate
目打：櫛型13½	perforation : comb perf. 13½
印面寸法：22.5×27mm	printing area size : 22.5×27mm
発行枚数：300万枚	number issued : 3 million
シート構成：20枚（横4×縦5）	pane format : 20 subjects (4×5)
製版・目打組合せ：タイプ1A-3	plate and perforation configuration : type 1A-3

記念・特殊切手（額面「円位」時期、ローマ字なし）

C237-238 皇太子（明仁）帰朝記念　発行日：1953（昭和28）.10.12
Return of Crown Prince Akihito from Abroad　Date of Issue：1953 (Showa 28).10.12

C237 鳳凰
Phoenix

C238 タンチョウ
Japanese Crane

●特印
Special Date Stamp

東京 Tokyo

C238 ●定常変種 Constant Flaws

A-pos.1

A-pos.17

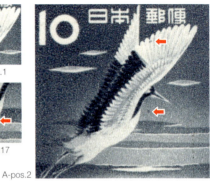
A-pos.2

	**	●	FDC
C237 5 yen くすみ赤 brown carmine	700	300	
C238 10 yen 暗い青 dark blue	1,500	500	
C237-238 (2種 set of 2 values)	2,200	800	3,000

切手データ Basic Information	
版式：C237=凹版速刷版、C238=板グラビア版	printing plate：C237= high-speed intaglio printing, C238=flat photogravure plate
目打：櫛型12½	perforation：comb perf. 12½
印面寸法：24×27mm	printing area size：24×27mm
発行枚数：各300万枚	number issued：3 million each
シート構成：20枚（横4×縦5）	pane format：20 subjects (4x5)
製版・目打組合せ：タイプ1A-3	plate and perforation configuration：type 1A-3

位置 position	特徴 feature	**	●
A-pos.1	翼に濃い青点、翼の下に青点 dark blue dot on wing, blue dot under wing	3,000	1,000
A-pos.2	翼に濃い青点、首の下に青点 dark blue dot on wing, blue dot below neck		
A-pos.17	胴体に青傷 blue spot on body		

C239-240 第8回国体記念　発行日：1953（昭和28）.10.22
8th National Athletic Meet　Date of Issue：1953 (Showa 28).10.22

C239 ラグビー Rugby

C240 柔道 Judo

●特印
Special Date Stamp

C239-240 松山
Matsuyama

	**	●	FDC
C239 5 yen 黒 black	1,200	400	
C240 5 yen こい緑 blue green	1,200	400	
C239-240 (2種連刷 pair)	3,700	2,000	4,000

切手データ Basic Information	
版式：板グラビア版	printing plate：flat photogravure plate
目打：櫛型13½	perforation：comb perf. 13½
印面寸法：27×22.5mm	printing area size：27×22.5mm
発行枚数：200万枚	number issued：2 million
シート構成：20枚（横5×縦4）	pane format：20 subjects (5x4)
製版・目打組合せ：タイプ1A-2	plate and perforation configuration：type 1A-2

C239-240 ●定常変種 Constant Flaws

A-pos.1

A-pos.14

C-pos.1

D-pos.5

A-pos.17

位置 position	特徴 feature	**	●
A-pos.1	"便"の右下に黒点 black dot to lower right of "便"	2,500	1,000
A-pos.14	足の左に濃い緑点 dark green dot to left of foot		
A-pos.17	太ももに黒点、"大"の上に汚れ black dot on thigh, smudge above "大"		
C-pos.1	足の下に大きい黒点 large black dot below foot		
D-pos.5	"5"の左に2個の点 two dots to left of "5"		

記念・特殊切手（額面「円位」時期、ローマ字なし）

C241 東京天文台創設75年記念
75th Anniversary of Tokyo Astronomical Observatory
発行日：1953（昭和28）.10.29　Date of Issue：1953 (Showa 28).10.29

C241 大赤道儀室と星座
Telescope Dome and Constellation

●特印 Special Date Stamp
東京 Tokyo

	**	●	FDC
C241 10 yen 暗い青 dark gray blue	2,200	550	3,500

切手データ Basic Information	
版式：板グラビア版	printing plate：flat photogravure plate
目打：櫛型13½	perforation：comb perf. 13½
印面寸法：27×22.5mm	printing area size：27×22.5mm
発行枚数：300万枚	number issued：3 million
シート構成：20枚（横5×縦4）	pane format：20 subjects (5x4)
製版・目打組合せ：タイプ1A-2	plate and perforation configuration：type 1A-2

C241 ●定常変種 Constant Flaws

A-pos.1　C-pos.15

D-pso.2　D-pos.5

C-pos.17

位置 position	特徴 feature	**	●
A-pos.1	第1コーナーに斜線 diagonal line in NW corner	5,000	1,500
C-pos.15	"0"の上に白抜けとインクの流れ white spot and ink flow above "0"		
C-pos.17	"1"の上方に白抜け、"0"の上に濃いシミ white spot above "1", dark stain above "0"		
D-pso.2	"日"の下方に大きい白点 large white dot below "日"		
D-pos.5	星に四角のシミ square stain under star		

C242 男子スピードスケート世界選手権記念
Men's World Speed Skating Championships
発行日：1954（昭和29）.1.16　Date of Issue：1954 (Showa 29).1.16

C242 スピードスケート
Speed Skating

●特印 Special Date Stamp
札幌 Sapporo

	**	●	FDC
C242 10 yen くすみ群青 blue	1,100	500	2,500

切手データ Basic Information	
版式：板グラビア版	printing plate：flat photogravure plate
目打：櫛型13½	perforation：comb perf. 13½
印面寸法：27×22.5mm	printing area size：27×22.5mm
発行枚数：200万枚	number issued：2 million
シート構成：20枚（横5×縦4）	pane format：20 subjects (5x4)
製版・目打組合せ：タイプ1A-2	plate and perforation configuration：type 1A-2

●年号の位置 Position of Year

pane A
右寄り shifted to right

pane B, C, D
正常 normal

C242 ●定常変種 Constant Flaws

A-pos.4　A-pos.20　B-pos.11　B-pos.12

C-pos.2　C-pos.16　D-pos.3　D-pos.17

位置 position	特徴 feature	**	●
A-pos.4	"日"の左に青点 blue dot to left of "日"	2,500	1,200
A-pos.20	"10"の右上方に濃い青点 dark blue dot to upper right of "10"		
B-pos.11	肘の右に大小の青点 large and small blue dots to right of elbow		
B-pos.12	太腿の右に濃い点 dark dot to right of thigh		
C-pos.2	帽子と頭の右に青点 blue dots on head and to right of head		
C-pos.16	背中の上に青点 blue dot above back		
D-pos.3	"0"に欠け defect in "0"		
D-pos.17	"1"の左に白抜け white spot to left of "1"		

記念・特殊切手（額面「円位」時期、ローマ字なし）

C243 日本国際見本市記念 / Japan International Trade Fair

発行日：1954（昭和29）.4.10　Date of Issue：1954 (Showa 29).4.10

C243 輸出品の象徴
Thread, Pearls, Gears, Buttons and Globe

●特印 Special Date Stamp
大阪 Osaka

	**	●	FDC
C243 10 yen 暗い紅 dark red	1,100	400	2,500

切手データ Basic Information	
版式：板グラビア版	printing plate : flat photogravure plate
目打：櫛型13½	perforation : comb perf. 13½
印面寸法：22.5×27mm	printing area size : 22.5×27mm
発行枚数：300万枚	number issued : 3 million
シート構成：20枚（横4×縦5）	pane format : 20 subjects (4x5)
製版・目打組合せ：タイプ1A-3	plate and perforation configuration : type 1A-3

C243 ●定常変種 Constant Flaws

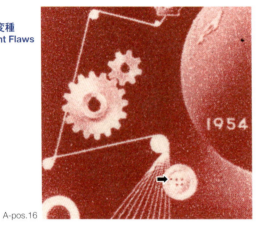
A-pos.16

A-pos.20

B-pos.2

B-pos.3

位置 position	特徴 feature	**	●
A-pos.16	ボタンの穴1個増 extra button hole	2,500	1,000
A-pos.20	"1"の上方に濃い赤点 dark red dot above "1"		
B-pos.2	"日"に小さな赤点 small dot on "日"		
B-pos.3	ボタンに赤点2個 two red dots on button		

C244 フリースタイルレスリング世界選手権記念 / World Freestyle Wrestling Championships

発行日：1954（昭和29）.5.22　Date of Issue：1954 (Showa 29).5.22

C244 フリースタイルレスリング
Freestyle Wrestling

●特印 Special Date Stamp
東京 Tokyo

	**	●	FDC
C244 10 yen 暗い黄味緑 deep green	800	250	2,500

切手データ Basic Information	
版式：凹版速刷版	printing plate : high-speed intaglio printing
目打：櫛型13½	perforation : comb perf. 13½
印面寸法：33×22.5mm	printing area size : 33×22.5mm
発行枚数：400万枚	number issued : 4 million
シート構成：20枚（横5×縦4）	pane format : 20 subjects (5x4)
製版・目打組合せ：タイプ1A-2	plate and perforation configuration : type 1A-2

C245-246 第9回国体記念 / 9th National Athletic Meet

発行日：1954（昭和29）.8.22　Date of Issue：1954 (Showa 29).8.22

C245 卓球 Table Tennis　C246 弓道 Kyudo (Japanese Archery)

●特印 Special Date Stamp
C245-246
札幌 Sapporo

	**	●	FDC
C245 5 yen 灰味茶 dull brown	1,000	350	
C246 5 yen 灰味オリーブ gray green	1,000	350	
C245-246 (2種連刷pair)	2,200	1,500	3,000

切手データ Basic Information	
版式：凹版速刷版	printing plate : high-speed intaglio printing
目打：櫛型12	perforation : comb perf. 12
印面寸法：27×27mm	printing area size : 27×27mm
発行枚数：各250万枚	number issued : 2.5 million each
シート構成：20枚（横4×縦5）	pane format : 20 subjects (4x5)
製版・目打組合せ：タイプ1K-1	plate and perforation configuration : type 1K-1

記念・特殊切手（額面「円位」時期、ローマ字なし）

C247-248 国際電気通信連合（ITU）加盟75周年記念
75th Anniversary of Admission to International Telecommunication Union (ITU)

発行日：1954（昭和29）.10.13　　Date of Issue：1954 (Showa 29).10.13

		**	●	FDC
C247	5 yen 暗い紫 dark purple-brown	550	200	
C248	10 yen 暗い青 deep blue	1,300	400	
C247-248	(2種 set of 2 values)	1,850	600	2,000

切手データ Basic Information

版式：凹版速刷版	printing plate：high-speed intaglio printing
目打：櫛型13½	perforation：comb perf. 13½
印面寸法：C247＝23×27mm、C248＝27×22.5mm	printing area size：C247＝23×27mm、C248＝27×22.5mm
発行枚数：各300万枚	number issued：3 million each
シート構成：C247＝20枚（横4×縦5）、C248＝20枚（横5×縦4）	pane format：C247＝20 subjects of (4×5), C248＝20 subjects (5×4)
製版・目打組合せ：C247＝タイプ1A-3、C248＝タイプ1A-2	plate and perforation configuration：C247＝type 1A-3, C248＝type 1A-2

C247 初期の電信機 Morse Telegraph Instrument

C248 ITU記念碑 ITU Monument

●特印 Special Date Stamp

東京 Tokyo

CP1／CB1 切手趣味週間
Philatelic Week 1954

発行日：1954（昭和29）.11.20　　Date of Issue：1954 (Showa 29).11.20

CP1 観音菩薩像（法隆寺）
Portrait of Bodhisattva (Kannon Bosatsu) (Horyuji Temple)

表紙 Booklet Cover
CB1　表面 front

●特印 Special Date Stamp
東京 Tokyo

	**	●	FDC
CP1 100 yen [10 yen (#358) ×10] くすみ赤紫・茶紅 dull reddish purple and rose-carmine	40,000	38,000	45,000
CB1		40,000	

切手データ Basic Information

版式：凸版輪転版	printing plate：rotary press relief printing
目打：全型13×13½、ペーンの上下目打なし	perforation：harrow perf. 13×13½, imperforate at top and bottom edges of pane
シート寸法：150×52mm	sheet size：150×52mm
発行枚数：65,000枚	number issued：65,000

タブの標語 tabs with postal slogans

CP1-pos.7「郵便は世界を結ぶ」"The mail connects the world"

CP1-pos.12「切手に学ぶ世界の知識」"Learn about the world from stamps"

●参考：普通切手の切手帳 Reference: Booklet of Defintives

BP29　観音菩薩像（法隆寺）
Portrait of Bodhisattva (Kannon Bosatsu) (Horyuji Temple)

表紙 Booklet Cover
B23　表面 front
(CB1と共通 common to CB1)

タブの標語 tabs with postal slogans

P29-pos.7「名あては正しく明りょうに」"Please write the address correctly and clearly"

BP29-pos.12「旅行に便利な簡易書簡」"Letter sheets are convenient for travel"

記念・特殊切手（額面「円位」時期、ローマ字なし）

CP1 ●定常変種 Constant Flaws

Pane 1-pos.2

Pane 2-pos.2　Pane 2-pos.3　
正常　白耳なりかけ　白耳
normal ear　partial ear contours　ear contours completely missing

位置 position	特徴 feature	**	●
Pane 1-pos.2	白目 white eye	100,000	80,000
Pane 2-pos.2	頭の左に十字形 cross to left of head		
Pane 2-pos.3	白耳 ear contours missing		

1) CP1は実用版1回転で10ペーン分が印刷され、10ペーンのうち2つのペーンに定常変種が確認されている。
1) There are ten panes in the plate; the constant flaws are known in two panes among the ten.

C249　第15回国際商業会議所総会記念
15th Congress of International Chamber of Commerce (ICC)

発行日：1955（昭和30）.5.16　　Date of Issue : 1955 (Showa 30) .5.16

C249 こいのぼり
Carp Streamer

●特印 Special Date Stamp

東京 Tokyo

		**	●	FDC
C249 10 yen 多色 multicolored		1,200	450	2,500

切手データ Basic Information	
版式：グラビア輪転版	printing plate : rotary photogravure plate
目打：櫛型13×13½	perforation : comb perf. 13×13½
印面寸法：31×27mm	printing area size : 31×27mm
発行枚数：300万枚	number issued : 3 million
シート構成：20枚（横4×縦5）	pane format : 20 subjects (4x5)
製版・目打組合せ：タイプ1J-2	plate and perforation configuration : type 1J-2

C249 ●定常変種 Constant Flaws

A-pos.2　A-pos.5　A-pos.13

B-pos.2　B-pos.10　B-pos.11

位置 position	特徴 feature	**	●
A-pos.2	"日"の左方に傷 defect to left of "日"	2,500	1,000
A-pos.5	"E"の先端が斜め slanted tip of stroke in "E"		
A-pos.13	"商業"の上方に白抜け white spot above "商業"		
B-pos.2	第2コーナーに白点 white dot in NE corner		
B-pos.10	"念"にかすれ scrape in "念"		
B-pos.11	"F"に黒点、"I"の先端欠け black dot in "F", slanted tip on "I"		

1) この切手は、多色刷グラビア輪転版による最初の記念切手で、製版・目打組合せタイプ1J-2で穿孔された。C249以降は刷色数に関わらず全て多色刷グラビア輪転版となり、スクリーンの線数は260線、スクリーン角度は刷色ごとに異なる。
1) This (C249) was the first commemorative stamp printed on the multicolor rotary photogravure press. The plate and perforation configuration is 1J-2. All photogravure stamps after C249 are printed on the multicolor rotary press regardless of the number of colors, the screen number being 260 lines. The screen angle differs for each color.

C250-251　第10回国体記念
10th National Athletic Meet

発行日：1955（昭和30）.10.30　　Date of Issue : 1955 (Showa 30) .10.30

C250 ロードレース　C251 マスゲーム
Road Race　Massed Gymnastics

●特印 Special Date Stamp

東京 Tokyo

	**	●	FDC
C250 5 yen 灰味青 bluish black	700	300	
C251 5 yen 暗い赤 brown-lake	700	300	
C250-251 (2種連刷 pair)	1,600	1,300	2,500

切手データ Basic Information	
版式：凹版速刷版	printing plate : high-speed intaglio printing
目打：櫛型13½	perforation : comb perf. 13½
印面寸法：22.5×27mm	printing area size : 22.5×27mm
発行枚数：各300万枚	number issued : 3 million each
シート構成：20枚（横4×縦5）	pane format : 20 subjects (4x5)
製版・目打組合せ：タイプ1A-3	plate and perforation configuration : type 1A-3

記念・特殊切手（額面「円位」時期、ローマ字なし）

C252　切手趣味週間
Philatelic Week 1955

発行日：1955（昭和30）.11.1
Date of Issue：1955 (Showa 30).11.1

C252 喜多川歌麿画「ビードロを吹く娘」
"Woman with a Glass Noisemaker (Poppen)" by Kitagawa Utamaro

●特印
Special Date Stamp

東京 Tokyo

	**	●	FDC
C252　10 yen 多色 multicolored	2,800	2,000	7,000

切手データ Basic Information

版式：グラビア輪転版	printing plate : rotary photogravure plate
目打：櫛型13½	perforation : comb perf. 13½
印面寸法：33×48mm	printing area size : 33×48mm
発行枚数：550万枚	number issued : 5.5 million
シート構成：10枚（横5×縦2）	pane format : 10 subjects (5x2)
製版・目打組合せ：タイプ1B-2	plate and perforation configuration : type 1B-2

C252 ●定常変種 Constant Flaws

D-pos.2

D-pos.5　　D-pos.8　　D-pos.10

位置 position	特徴 feature	**	●
A-pos.3	上耳紙に黒点群 black dots on top selvage	6,000	4,000
A-pos.9	顎にホクロ black dot on chin		
B-pos.1	"日"の上に黒点群 black dots above "日"		
B-pos.4	黒点群と目の下に黒点 black dots, and black dot under eye		
B-pos.5	目の下に黒点 black dot under eye		
C-pos.3	目の上と髷の下に黒点 black dots above eye and under topknot		
C-pos.4	髷の下に黒点群 black dots under topknot		
C-pos.6	首に黒点群 black dots on neck		
D-pos.2	黒点群3ヵ所 three black dots		
D-pos.5	ビードロの左方に黒点 black dot to left of glass poppen		
D-pos.8	ビードロの上に黒点 black dot above glass poppen		
D-pos.10	右腕に白抜き white mark on right arm		

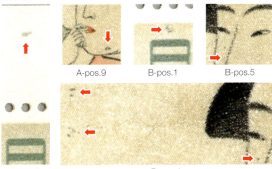

A-pos.9　　B-pos.1　　B-pos.5

A-pos.3　　　　　B-pos.4

C-pos.3　　C-pos.4　　C-pos.6

1）この時期の実用版構成は2×2面頭合せで製版するのが一般的であるが、この切手は2×2面順並びで製版されている。以後の趣味週間切手も同様である。

1) Although the plate configuration was usually 2×2 panes head-to-head (1A-2) at this period, for this stamp it was 2×2 panes normal (1B-2), the same as with the subsequent "Philatelic Week" stamps.

C253　第23回世界卓球選手権大会記念
23rd World Table Tennis Championships

発行日：1956（昭和31）.4.2
Date of Issue：1956 (Showa 31).4.2

C253 卓球 Table Tennis

●特印
Special Date Stamp

東京 Tokyo

	**	●	FDC
C253　10 yen 赤茶 red brown	340	200	1,000

切手データ Basic Information

版式：グラビア輪転版	printing plate : rotary photogravure plate
目打：櫛型13½	perforation : comb perf. 13½
印面寸法：22.5×27mm	printing area size : 22.5×27mm
発行枚数：500万枚	number issued : 5 million
シート構成：20枚（横4×縦5）	pane format : 20 subjects (4x5)
製版・目打組合せ：タイプ1A-3	plate and perforation configuration : type 1A-3

記念・特殊切手（額面「円位」時期、ローマ字なし）

C253 ●定常変種 Constant Flaws

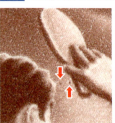
A-pos.1

A-pos.5

A-pos.8

B-pos.8

C-pos.1

D-pos.12　D-pos.19

C-pos.12

D-pos.11

位置 position	特徴 feature	**	●
A-pos.1	ラケット近くに茶点2ヵ所 two brown dots near racket	800	400
A-pos.5	ユニホームに濃い点 dark dot on uniform		
A-pos.8	"本"の左に白点 white dot to left of "本"		
B-pos.8	腕に大きい茶点 large brown dot on arm		
C-pos.1	目の近くに白ぼやけ white blur near eye		
C-pos.12	腕の下に茶点 brown dot under arm		
D-pos.11	ネットに茶点 brown dot on net		
D-pos.12	"界"の右下に茶点 brown dot to lower right of "界"		
D-pos.19	左マージンに茶点 brown dot in left margin		

C254 世界柔道選手権大会記念　1st World Judo Championships

発行日：1956（昭和31）.5.3　Date of Issue：1956 (Showa 31) .5.3

C254 柔道
Judo

●特印
Special Date Stamp

東京 Tokyo

C254　10 yen 暗い赤紫・濃い緑 green and lilac　** 400　● 200　FDC 1,500

切手データ Basic Information	
版式：凹版速刷版	printing plate：high-speed intaglio printing
目打：櫛型13½	perforation：comb perf. 13½
印面寸法：27×22.5mm	printing area size：27×22.5mm
発行枚数：500万枚	number issued：5 million
シート構成：20枚（横5×縦4）	pane format：20 subjects (5x4)
製版・目打組合せ：タイプ1A-2	plate and perforation configuration：type 1A-2

C255 世界こどもの日制定記念　Establishment of World Children's Day

発行日：1956（昭和31）.5.5　Date of Issue：1956 (Showa 31) .5.5

C255 子供とこいのぼり
Children and Carp Streamer

●特印
Special Date Stamp

東京 Tokyo

切手データ Basic Information	
版式：グラビア輪転版	printing plate：rotary photogravure plate
目打：櫛型13×13½	perforation：comb perf. 13×13½
印面寸法：40×22.5mm	printing area size：40×22.5mm
発行枚数：600万枚	number issued：6 million
シート構成：20枚（横5×縦4）	pane format：20 subjects (5x4)
製版・目打組合せ：タイプ1E-1	plate and perforation configuration：type 1E-1

C255　10 yen 灰黒・明るい緑青 black and light blue　** 300　● 150　FDC 1,000

C255 ●定常変種 Constant Flaws

A-pos.1

A-pos.18

A-pos.19

記念・特殊切手（額面「円位」時期、ローマ字なし）

B-pos.4

B-pos.14

B-pos.17

D-pos.18

D-pos.19

C-pos.8

C-pos.16

C-pos.19

D-pos.17

位置 position	特徴 feature	**	●
A-pos.1	鯉のぼりの口に黒点 black dot on mouth of carp streamer	600	300
A-pos.18	鯉のぼりの尾上方にリタッチ retouch above tail of carp streamer		
A-pos.19	"記"と"念"の間に白抜け white spot between "記" and "念"		
B-pos.4	"郵"に黒点 black dot on "郵"		
B-pos.14	服に黒点2個 two black dots on clothing		
B-pos.17	指先の左下方に青のリタッチ blue retouch below left of fingertip		
C-pos.8	白ぼやけ2ヵ所 two white blurs		
C-pos.16	"界"の上に白抜け white spot above "界"		
C-pos.19	"造"(銘版)の右下方に黒点 black dot below right of "造" (imprint)		
D-pos.17	白抜けと黒点2ヵ所 white spot and two black dots		
D-pos.18	脇の下に黒点 black dot under armpit		
D-pos.19	黒点2ヵ所 two black dots		

C256 東京開都500年記念　Tokyo Quincentenary
発行日：1956（昭和31）.10.1
Date of Issue：1956 (Showa 31).10.1

C256 皇居周辺
Palace Moat and Modern Tokyo

●特印
Special Date Stamp

東京 Tokyo

C256　10 yen 暗い茶緑 dull purple ………… 500　　200　2,000
　　　　　　　　　　　　　　　　　　　　　　　**　●　FDC

切手データ Basic Information	
版式：凹版速刷版	printing plate : high-speed intaglio printing
目打：櫛型13×13½	perforation : comb perf. 13×13½
印面寸法：40×22.5mm	printing area size : 40×22.5mm
発行枚数：500万枚	number issued : 5 million
シート構成：20枚（横5×縦4）	pane format : 20 subjects (5x4)
製版・目打組合せ：タイプ1G	plate and perforation configuration : type 1G

C257 佐久間ダム竣工記念　Completion of Sakuma Dam
発行日：1956（昭和31）.10.15
Date of Issue：1956 (Showa 31).10.15

C257 佐久間ダム
Sakuma Dam

●特印
Special Date Stamp

静岡中部
Shizuoka Nakabe

C257　10 yen 暗い青 dark blue ………… 500　　200　1,300
　　　　　　　　　　　　　　　　　　　　　　**　●　FDC

切手データ Basic Information	
版式：凹版速刷版	printing plate : high-speed intaglio printing
目打：櫛型13½	perforation : comb perf. 13½
印面寸法：22.5×33mm	printing area size : 22.5×33mm
発行枚数：500万枚	number issued : 5 million
シート構成：20枚（横4×縦5）	pane format : 20 subjects (4x5)
製版・目打組合せ：タイプ1A-3	plate and perforation configuration : type 1A-3

記念・特殊切手（額面「円位」時期、ローマ字なし）

C258-259 第11回国体記念
11th National Athletic Meet

発行日：1956（昭和31）.10.28
Date of Issue : 1956 (Showa 31).10.28

C258 走り幅跳び Long Jump
C259 バスケットボール Basketball

●特印
Special Date Stamp

西宮 Nishinomiya

			**	●	FDC
C258	5 yen 暗い紫茶 brown violet	………	280	150	
C259	5 yen 暗い緑青 steel blue	………	280	150	
	C258-259（2種連刷 pair）	………	600	650	1,000

切手データ Basic Information	
版式：凹版速刷版	Printing Plate : high-speed intaglio printing
目打：櫛型13½	Perforation : Comb 13½
印面寸法：22.5×27.5mm	printing area size : 22.5×27.5mm
発行枚数：各400万枚	Number Issued : 4 million each
シート構成：20枚（横4×縦5）	Pane Format : 20 subjects (4x5)
製版・目打組合せ：タイプ1A-3	plate and perforation configuration : type 1A-3

C260 切手趣味週間
Philatelic Week 1956

発行日：1956（昭和31）.11.1
Date of Issue : 1956 (Showa 31).11.1

C260 東洲斎写楽画「市川蝦蔵」"Kabuki Actor Ichikawa Ebizo" by Toshusai Sharaku

●特印
Special Date Stamp

東京 Tokyo

	**	●	FDC
C260 10yen 多色 multicolored	……… 2,400	1,800	6,500

切手データ Basic Information	
版式：グラビア輪転版	printing plate : rotary photogravure plate
目打：櫛型13½	perforation : comb perf.13½
印面寸法：33×48mm	printing area size : 33×48mm
発行枚数：550万枚	number issued : 5.5 million each
シート構成：10枚（横5×縦2）	pane format : 10 subjects (5x2)
製版・目打組合せ：タイプ1B-2	plate and perforation configuration : type 1B-2

●スクリーン角度 Screen Angles

1版 Plate 1 橙色59度 orange 59 degrees

2版 plate 2 橙色74度 orange 74 degrees

C260 ●定常変種 Constant Flaws

P1/P2-A-pos.1

P1-A-pos.5

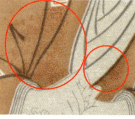
P2-A-pos.2

P1/P2-B-pos.1

P2-B-pos.2

P1/P2-C-pos.2

P2-C-pos.7　　P1/P2-D-pos.2

P1-C-POS.8

P2-D-pos.2

P1-D-pos.8

位置 position	特徴 feature	**	●
P1/P2-A-pos.1	"便"の左に白点 white dot to left of "便"	6,000	4,000
P1-A-pos.5	耳たぶの上部に黒点 black dot on earlobe		
P2-A-pos.2	濡れ衣（水濡れ状の濃いシミ） dark "water stain" on clothing		
P1/P2-B-pos.1	鼻の左と上唇の上に黒点 black dots to left of nose and above upper lip		
P2-B-pos.2	頭の左に黒と白点群 black and white dots to left of head		
P1/P2-C-pos.2	家紋の左に丸いシミと白点 round stain and white dot to left of family crest		
P1-C-pos.8	半襟に小さい黒点2個 small two black dots on Han-eri collar		
P2-C-pos.7	"0"の上にシミ stain above "0"		
P1/P2-D-pos.2	頭の左に白抜け white spot to left of head		
P2-D-pos.2	"本"の右方に白点、頭の左に黒点 white dot to right of "本", black dot to left of head		
P1-D-pos.8	顎と首に小さい黒点 small black dot on chin and neck		

記念・特殊切手（額面「円位」時期、ローマ字なし）

C261 マナスル登頂記念　発行日：1956（昭和31）.11.3
Conquest of Manaslu　Date of Issue: 1956 (Showa 31).11.3

C261 マナスルの遠望
Mount Manaslu

●特印
Special Date Stamp

東京 Tokyo

	**	●	FDC
C261　10 yen 多色 multicolored	800	450	3,000

切手データ Basic Information

版式：グラビア輪転版	printing plate : rotary photogravure plate
目打：櫛型13×13½	perforation : comb perf. 13×13½
印面寸法：40×22.5mm	printing area size : 40×22.5mm
発行枚数：500万枚	number issued : 5 million
シート構成：20枚（横5×縦4）	pane format : 20 subjects (5×4)
製版・目打組合せ：タイプ1E-1	plate and perforation configuration : type 1E-1

C261 ●定常変種 Constant Flaws

A-pos.12　　A-pos.13　　B-pos.1　　B-pos.19

C-pos.1　　C-pos.3

位置 position	特徴 feature	**	●
A-pos.12	山肌に青点 blue dot on mountain surface	2,000	1,000
A-pos.13	"5"の上方に白点2個、ザイルの右に白点 two white dots above "5", and white dot to right of climbing rope		
B-pos.1	頭の左に白線 white line to left of head		
B-pos.19	山肌に白点と半円状の白線 white dot and white line on mountain surface		
C-pos.1	山頂の右に斜線と白点 diagonal line and white dots to right of mountaintop		
C-pos.3	帽子に茶点、頭の右に白点 brown dot on hat, white dot to right of head		

C262 東海道電化完成記念　発行日：1956（昭和31）.11.19
Electrification of Tokaido Line　Date of Issue: 1956 (Showa 31).11.19

C262 EF58形電気機関車と歌川広重画
「東海道五拾三次之内 由井」
EF58 Electric Locomotive and "The Fifty-three Stations of the Tokaido: Yui" by Utagawa Hiroshige

●特印
Special Date Stamp

東京 Tokyo

	**	●	FDC
C262　10 yen 多色 multicolored	1,200	450	3,000

切手データ Basic Information

版式：グラビア輪転版	printing plate : rotary photogravure plate
目打：櫛型13×13½	perforation : comb perf. 13×13½
印面寸法：40×22.5mm	printing area size : 40×22.5mm
発行枚数：500万枚	number issued : 5 million
シート構成：20枚（横5×縦4）	pane format : 20 subjects (5×4)
製版・目打組合せ：タイプ1E-1	plate and perforation configuration : type 1E-1

C262 ●定常変種 Constant Flaws

A-pos.10

B-pos.14

B-pos.15　　B-pos.3

位置 position	特徴 feature	**	●
A-pos.10	富士山の上方に茶点 brown dot above Mt. Fuji	2,500	1,000
B-pos.3	機関車に黒点 black dot on locomotive		
B-pos.14	帆船の左に茶点 brown dot to left of sailing ship		
B-pos.15	斜面に茶点 brown dot on spole		

記念・特殊切手（額面「円位」時期、ローマ字なし）

C263 日本機械巡航見本市記念 / Japan Machinery Floating Fair

発行日：1956（昭和31）.12.18
Date of Issue：1956 (Showa 31).12.18

C263 見本市船「日昌丸」と真空管に歯車
Cogwheel, Vacuum Tube and Floating Fair Ship "Nissho Maru"

●特印 Special Date Stamp

東京 Tokyo

		**	●	FDC
C263	10 yen 暗い青 red brown	150	180	900

切手データ Basic Information

版式：凹版速刷版	printing plate：high-speed intaglio printing
目打：櫛型13½	perforation：comb perf. 13½
印面寸法：33×22.5mm	printing area size：33×22.5mm
発行枚数：500万枚	number issued：5 million
シート構成：20枚（横5×縦4）	pane format：20 subjects (5x4)
製版・目打組合せ：タイプ1A-2	plate and perforation configuration：type 1A-2

C264 国際連合加盟記念 / Admission to United Nations

発行日：1957（昭和32）.3.8
Date of Issue：1957 (Showa 32).3.8

C264 国連マーク
United Nations Emblem

●特印 Special Date Stamp

東京 Tokyo

		**	●	FDC
C264	10 yen 青・紫赤 light blue and red brown	180	150	800

切手データ Basic Information

版式：グラビア輪転版・凹版速刷版	printing plate：rotary photogravure plate & high-speed intaglio printing
目打：櫛型13½	perforation：comb perf. 13½
印面寸法：33×22.5mm	printing area size：33×22.5mm
発行枚数：500万枚	number issued：5 million
シート構成：20枚（横5×縦4）	pane format：20 subjects (5x4)
製版・目打組合せ：タイプ1B-2	plate and perforation configuration：type 1B-2

C264 ●定常変種 Constant Flaws

 A-pos.10
 A-pos.14
 B-pos.2
 B-pos.14
 C-pos.3
 C-pos.14
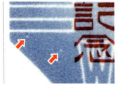 D-pos.2
 D-pos.18

位置 position	特徴 feature	**	●
A-pos.10	"日"の上方に白点 white dot above "日"	400	200
A-pos.14	"S"に青点 blue dot on "S"		
B-pos.2	マークに大きな青点 large blue dot on emblem		
B-pos.14	"O"の右に白点、右下に白線 white dot to right of "O"; white line below right of "O"		
C-pos.3	"O"に円弧状の白線 circular white line on "O"		
C-pos.14	"I"と"C"の間に大きな青点 large blue dot between "I" and "C"		
D-pos.2	印面左下に白点2個 two white dots in lower left printing area		
D-pos.18	"O"と"E"に白点 white dots in "O" and "E"		

1) この時期の実用版構成は2×2面頭合せで製版するのが一般的であるが、この切手は2×2面順並びで製版されている。

1) Although the plate configuration was usually 2×2 panes head-to-head (1A-2) at this period, for this stamp it was 2×2 panes normal (1B-2).

記念・特殊切手（額面「円位」時期、ローマ字なし）

C265 国際地球観測年（IGY）記念　発行日：1957（昭和32）.7.1
International Geophysical Year　Date of Issue：1957 (Showa 32).7.1

C265 IGYマークに南極観測船「宗谷」とコウテイペンギン
IGY Emblem, Emperor Penguin and Ship "Soya"

●特印 Special Date Stamp

東京 Tokyo

	**	●	FDC
C265　10 yen 青・黒・黄 blue, black and yellow	150	120	800

切手データ Basic Information

版式：グラビア輪転版	printing plate : rotary photogravure plate
目打：櫛型13½	perforation : comb perf. 13½
印面寸法：22.5×27mm	printing area size : 22.5×27mm
発行枚数：600万枚	number issued : 6 million
シート構成：20枚（横4×縦5）	pane format : 20 subjects (4×5)
製版・目打組合せ：タイプ1A-3	plate and perforation configuration : type 1A-3

C265 ●定常変種 Constant Flaws

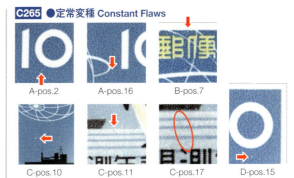

A-pos.2　A-pos.16　B-pos.7
C-pos.10　C-pos.11　C-pos.17　D-pos.15

位置 position	特徴 feature	**	●
A-pos.2	"0"の下に白点 white dot under "0"	400	200
A-pos.16	"1"の左下に白点 white dot below left of "1"		
B-pos.7	"便"の上に白点 white spot above "便"		
C-pos.10	地球儀の左下方に白抜け white spot below left of globe		
C-pos.11	船の右下に青点 blue dot below right of ship		
C-pos.17	船と"測"の間に青い斜線 blue diagonal line between ship and "測"		
D-pos.15	"0"の下に白点 white dot under "0"		

C266 原子炉竣工記念　発行日：1957（昭和32）.9.18
Completion of Nuclear Reactor　Date of Issue：1957 (Showa 32).9.18

C266 原子炉 Nuclear Reactor

●特印 Special Date Stamp

東京 Tokyo

	**	●	FDC
C266　10 yen 暗い紫 dark purple	80	60	650

切手データ Basic Information

版式：凹版速刷版	printing plate : high-speed intaglio printing
目打：櫛型13½	perforation : comb perf. 13½
印面寸法：23×27mm	printing area size : 23×27mm
発行枚数：700万枚	number issued : 7 million
シート構成：20枚（横4×縦5）	pane format : 20 subjects (4×5)
製版・目打組合せ：タイプ1A-3	plate and perforation configuration : type 1A-3

C267-268 第12回国体記念　発行日：1957（昭和32）.10.26
12th National Athletic Meet　Date of Issue：1957 (Showa 32).10.26

C267 段違い平行棒 Uneven Bars　C268 ボクシング Boxing

●特印 Special Date Stamp

静岡 Shizuoka

	**	●	FDC
C267　5 yen 青 blue	50	50	
C268　5 yen 赤茶 dark red	50	50	
C267-268（2種連刷 pair）	120	200	500

切手データ Basic Information

版式：凹版速刷版	printing plate : high-speed intaglio printing
目打：櫛型13½	perforation : comb perf. 13½
印面寸法：22.5×27mm	printing area size : 22.5×27mm
発行枚数：各600万枚	number issued : 6 million each
シート構成：20枚（横4×縦5）	pane format : 20 subjects (4×5)
製版・目打組合せ：タイプ1A-3	plate and perforation configuration : type 1A-3

記念・特殊切手（額面「円位」時期、ローマ字なし）

C269 切手趣味週間
Philatelic Week 1957

発行日：1957（昭和32）.11.1
Date of Issue：1957（Showa 32）.11.1

C269 鈴木春信画
「まりつき」
"Girl Bouncing Ball"
by Suzuki Harunobu

●特印
Special Date Stamp

東京 Tokyo

	**	●	FDC
C269 10yen 多色 multicolored	340	340	2,800

C269 ●定常変種 Constant Flaws

A-pos.3 　　　　　　B-pos.3

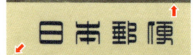

B-pos.5　　　　　　C-pos.4

C-pos.10　　D-pos.7　　D-pos.9

切手データ Basic Information

版式：グラビア輪転版	printing plate : rotary photogravure plate
目打：櫛型13½	perforation : comb perf. 13½
印面寸法：33×48mm	printing area size：33×48mm
発行枚数：850万枚	number issued：8.5 million
シート構成：10枚（横5×縦2）	pane format : 10 subjects (5×2)
製版・目打組合せ：タイプ1B-2	plate and perforation configuration : type 1B-2

●色調 Color Shades

1刷 1st printing
背景：明るい
background : bright

不明
printing unknown
背景：赤味やや強い
background : slightly reddish

増刷
additional printing
背景：緑味やや強い
background : slightly greenish

1）1刷の800万枚が即日完売したため、急遽50万枚を増刷した。増刷は1刷と比べてやや濃く緑がかっている。ただし、1刷にも色が濃いものがあるため、判別は難しい。
1）After the first printing of 8 million stamps was sold out on the day of issue, an additional printing of 500,000 more stamps was quickly produced. The background color of the additional-printing stamps is slightly darker and greenish in comparison with that of the first-printing stamps. However, some of the first-printing stamps have dark colors also, so it is difficult to distinguish them.

位置 position	特徴 feature	**	●
A-pos.3	背中の左方に青点、指に線欠け blue dot to left of back, missing line on finger	800	800
B-pos.3	第4コーナーに黒点2個、帯に青点 two black dots in SW corner, blue dot on obi		
B-pos.5	"日"の左下方に大きな白点、"便"の上方に黒点 large white dot below left of "日", black dot above "便"		
C-pos.4	第1コーナーに黒点 black dot in NW corner		
C-pos.10	第1コーナーに黒点2個、指の右に青点 two black dots in NW corner, blue dot to right of finger		
D-pos.7	"便"の上と下に白点 white dots above and under "便"		
D-pos.9	帯に青点と白点 blue and white dots on obi		

C270 小河内（おごうち）ダム竣工記念
Completion of Ogouchi Dam

発行日：1957（昭和32）.11.26
Date of Issue：1957（Showa 32）.11.26

C270 奥多摩湖と小河内ダム Lake Okutama and Ogouchi Dam

●特印
Special Date Stamp

小河内 Ogouchi

	**	●	FDC
C270 10yen 暗い青 ultramarine	80	50	800

切手データ Basic Information

版式：凹版速印版	printing plate : high-speed intaglio printing
目打：櫛型13½	perforation : comb perf. 13½
印面寸法：27×22.5mm	printing area size：27×22.5mm
発行枚数：800万枚	number issued：8 million
シート構成：20枚（横5×縦4）	pane format : 20 subjects (5×4)
製版・目打組合せ：タイプ1A-2	plate and perforation configuration : type 1A-2

C270 ●定常変種 Constant Flaws

C-pos.4

位置 position	特徴 feature	**	●
C-pos.4	"1"のセリフ欠け part of serif missing on "1"	200	100

記念・特殊切手（額面「円位」時期、ローマ字なし）

C271 製鉄100年記念　発行日：1957（昭和32）.12.1
Iron Manufacturing Industry Centenary　Date of Issue：1957(Showa 32).12.1

C271 日本最初の高炉と1957年当時の溶鉱炉 Early and Modern Japanese Blast Furnaces

●特印
Special Date Stamp

東京 Tokyo

C271 ●定常変種 Constant Flaws

A-pos.3　A-pos.19　B-pos.15　B-pos.20

C-pos.8　C-pos.12/17　D-pos.8　D-pos.9

	**	●	FDC
C271　10 yen だいだい・暗い紫 orange and dark purple	50	50	800

切手データ Basic Information	
版式：グラビア輪転版	printing plate：rotary photogravure plate
目打：櫛型13½	perforation：comb perf. 13½
印面寸法：33×22.5mm	printing area size：33×22.5mm
発行枚数：800万枚	number issued：8 million
シート構成：20枚（横5×縦4）	pane format：20 subjects (5x4)
製版・目打組合せ：タイプ1A-2	plate and perforation configuration：type 1A-2

位置 position	特徴 feature	**	●
A-pos.3	橙色点 orange dot	200	200
A-pos.19	橙色の点群 orange dots		
B-pos.15	橙色点 orange dot		
B-pos.20	橙色の太線 thick orange line		
C-pos.8	橙色の太点 thick orange dot		
C-pos.12/17	橙色の太点2ヵ所 two thick orange dots		
D-pos.8	紫点 purple dot		
D-pos.9	橙色の大きな点 large orange spot		

C272 関門トンネル開通記念　発行日：1958（昭和33）.3.9
Opening of Kanmon Tunnel　Date of Issue：11958(Showa 33).3.9

C272 関門トンネルの断面図 Cross Section of Kanmon Tunnel

●特印
Special Date Stamp

門司 Moji

	**	●	FDC
C272　10 yen 多色 multicolored	50	50	800

切手データ Basic Information	
版式：グラビア輪転版	printing plate：rotary photogravure plate
目打：櫛型13×13½	perforation：comb perf. 13×13½
印面寸法：40×22.5mm	printing area size：40×22.5mm
発行枚数：1,000万枚	number issued：10 million
シート構成：20枚（横5×縦4）	pane format：20 subjects (5x4)
製版・目打組合せ：タイプ1E-1	plate and perforation configuration：type 1E-1

A-pos.12　B-pos.2　B-pos.4　B-pos.15

C-pos.8　D-pos.8　D-pos.19

C272 ●定常変種 Constant Flaws

A-pos.5

位置 position	特徴 feature	**	●
A-pos.5	"郵"の上方に黒点、"1"に白点群と黒点 black dot above "郵", white dots and black dot near "1"	200	200
A-pos.12	"1"の左下に濃い斜線 dark line below left of "1"		
B-pos.2	車の右下に緑点 green dot below right of car		
B-pos.4	女性の頭上に白点 white dot above woman's head		
B-pos.15	"便"の右に濃い青点 dark blue dot to right of "便"		
C-pos.8	車の右下に緑点 green dot below right of car		
D-pos.8	濃い青点と黒点2個 dark blue dot and two black cots		
D-pos.19	"1"の上方に帯状の濃点群 dark band-like group of dots above "1"		

記念・特殊切手（額面「円位」時期、ローマ字なし）

C273 切手趣味週間
Philatelic Week 1958

発行日：1958（昭和33）.4.20
Date of Issue：1958 (Showa 33) .4.20

C273 鳥居清長画
「雨中湯帰り」
"Beauties Returning from Bath" by Torii Kiyonaga

●特印
Special Date Stamp

東京 Tokyo

		**	●	FDC
C273	10 yen 多色 multicolored	80	50	600

切手データ Basic Information	
版式：グラビア輪転版	printing plate : rotary photogravure plate
目打：櫛型13½	perforation : comb perf. 13½
印面寸法：33×48mm	printing area size : 33×48mm
発行枚数：2,500万枚	number issued : 25 million
シート構成：10枚（横5×縦2）	pane format : 10 subjects (5x2)
製版・目打組合せ：タイプ1B-2、1C-2	plate and perforation configuration : type 1B-2 and 1C-2

C273 ●定常変種 Constant Flaws

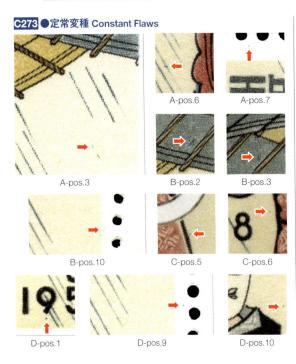

位置 position	特徴 feature		**	●
A-pos.3	右の傘の下に黒点 black dot below right of umbrella		200	100
A-pos.6	着物の左に緑点 green dot to left side of kimono			
A-pos.7	マージンに黒点 black dot in margin			
B-pos.2	濃い大きい緑点 large dark-green dot			
B-pos.3	濃い大きい緑点 large dark-green dot			
B-pos.10	マージンに緑点群 green dots in margin			
C-pos.5	"0"の下に黒点 black dot under "0"			
C-pos.6	"8"の右方に緑点 green dot to right of "8"			
D-pos.1	"9"の下に黒点 black dot under "9"			
D-pos.9	マージンに黒点 black dot in margin			
D-pos.10	着物の上方に黒点 black dot above kimono			

1) 目打穿孔には、2面縦並びのタイプ1B-2と4面田型のタイプ1C-2がある。タイプ1C-2は切手収集ブームを背景とした大幅な発行枚数増加に対応するため、この切手で初めて採用された。

1) There are two types of plate and perforation configuration for this stamp: 1B-2 and 1C-2. This stamp was the first instance of Type 1C-2, which was created in response to the drastically increased demand caused by a stamp-collecting boom.

C274 日本開港100年記念
Centenary of Opening of Ports in Japan

発行日：1958（昭和33）.5.10
Date of Issue：1958 (Showa 33) .5.10

C274 井伊直弼銅像と黒船
Statue of Ii Naosuke and Arrival of Perry

●特印
Special Date Stamp

横浜 Yokohama

		**	●	FDC
C274	10 yen 暗い赤・暗い青緑 gray blue and carmine	50	50	400

切手データ Basic Information	
版式：凹版速刷版	printing plate : high-speed intaglio printing
目打：櫛型13	perforation : Comb 13
印面寸法：35.5×25mm	printing area size : 35.5×25mm
発行枚数：1,500万枚	number issued : 15 million
シート構成：20枚（横5×縦4）	pane format : 20 subjects (5x4)
製版・目打組合せ：タイプ1A-3	plate and perforation configuration : type 1A-3

記念・特殊切手（額面「円位」時期、ローマ字なし）

C275-278　第3回アジア競技大会記念　発行日：1958（昭和33）.5.24
3rd Asian Games, Tokyo　Date of Issue : 1958 (Showa 33) .5.24

C275 国立競技場　　C276 聖火とマーク　　C277 ランナー　　C278 飛び込み　　特印 Special Date Stamp
National Stadium　　Torch and Emblem　　Runner　　Woman Diver　　東京 Tokyo

		**	●	FDC
C275	5 yen 多色 multicolored	40	40	
C276	10 yen 多色 multicolored	60	80	
C277	14 yen 多色 multicolored	60	60	
C278	24 yen 多色 multicolored	80	80	
	C275-278 (4種 set of 4 values)	240	260	1,000

切手データ Basic Information	
版式：グラビア輪転版	printing plate : rotary photogravure plate
目打：櫛型13½	perforation : comb perf. 13½
印面寸法：27×27mm	printing area size : 27×27mm
発行枚数：C275-276=各3,000万枚、C277-278=各1,000万枚	number issued : C275-276=30 million each, C277-278=10 million each
シート構成：各20枚（横4×縦5）	pane format : 20 subjects (4x5) each
製版・目打組合せ：タイプ1A-3	plate and perforation configuration : type 1A-3

C275 ●定常変種 Constant Flaws

A-pos.9　　A-pos.19　　B-pos.2　　B-pos.16

C-pos.3　　C-pos.19　　D-pos.16　　D-pos.20

位置 position	特徴 feature	**	●
A-pos.9	スタンド上段に黒点2個 two black dots on stadium seats	100	100
A-pos.19	トラックに緑点 green dot on track		
B-pos.2	左マージンに緑点 green dot in left margin		
B-pos.16	"5"に黒点 black dot on "5"		
C-pos.3	トラックに大きな濃い点 large dark dot on track		
C-pos.19	競技場の左に白線 white line to left of stadium		
D-pos.16	"A"の右に白点、マージンに大きな黒点2個 white dot to right of "A", two large black dots in margin		
D-pos.20	スタンドに小さい黒点 small black dot on stadium seats		

C276 ●定常変種 Constant Flaws

A-pos.7　　A-pos.11

B-pos.19　　C-pos.5

C-pos.12　　D-pos.7

B-pos.7

D-pos.19

位置 position	特徴 feature	**	●
A-pos.7	"A"の下に白線 white line under "A"	200	200
A-pos.11	トーチの左に濃い点 dark dot to left of torch		
B-pos.7	"RD"に太い線、炎の左に白点、3に赤点 thick line at "RD", white dot to left of flame, red dot at "3"		
B-pos.19	"N"の右に大きな赤点 large red dot to right of "N"		
C-pos.5	"0"の右下に赤点 red dot under right of "0"		
C-pos.12	"8"の右方に赤点 red dot to right of "8"		
D-pos.7	"日"の右に白抜け white dot to right of "日"		
D-pos.19	"A"の中に茶点、"5"の右に赤点 brown dot in "A", red dot to right of "5"		

●C277の赤色スクリーンによる版分類
Classification by Red Screen for C277

1版 Plate1　　右上がり upward-sloping
赤2度 red 2 degrees

2版 Plate2　　右下がり downward-sloping
赤84度 red 84 degrees

1) 赤色のスクリーン角度は2度と84度の2種が確認されている。スクリーン角度は、印面端のギザギザの傾斜で区別できる。

1) Although two types of photogravure screen angle (2 degrees and 84 degrees) are known for the red plate.

C277 ●定常変種 Constant Flaws

P1-A-pos.16　　P1/P2-B-pos.19　　P2-B-pos.19　　P1/P2-C-pos.5

P1/P2-C-pos.19　　P1-D-pos.2　　P1-D-pos.7　　P2-D-pos.7

位置 position	特徴 feature	**	●
P1-A-pos.16	第2コーナーに白点群 white dots in NE corner	200	200
P1/P2-B-pos.19	下マージンに黒線 black line in lower margin		
P2-B-pos.19	上マージンに赤点 red dot in upper margin		
P1/P2-C-pos.5	もみあげに赤点 red dot on sideburns		
P1/P2-C-pos.19	左腕に赤点 red dot on left arm		
P1-D-pos.2	ランニングパンツに赤点 red dot on running shorts		

右上につづく Continued to up right ↗

| P1-D-pos.7 | 目に黒点 black dot on eye | | |
| P2-D-pos.7 | 左腕に赤点 red dot on left arm | | |

C278 ●定常変種 Constant Flaws

A-pos.7　　B-pos.4　　B-pos.8

C-pos.19

D-pos.17

位置 position	特徴 feature	**	●
A-pos.7	白点3ヵ所 three white dots	200	200
B-pos.4	脚の下方に白点 white dot below legs		
B-pos.8	マークの右に白点3個、"本"の上方に白点2ヵ所 three white dots to right of emblem, two white dots above "本"		
C-pos.2	白点3ヵ所 three white dots		
C-pos.19	"第"の上方に白点 white dot above "第"		
D-pos.17	"ア"の上に白点群 white dots above "ア"		

C279　ブラジル移住50年記念　50th Anniversary of Japanese Emigration to Brazil

発行日：1958（昭和33）.6.18　Date of Issue：1958 (Showa 33).6.18

C279「笠戸丸」と南米地図
Kasato Maru, Map and Brazilian Flag Emblem

●特印 Special Date Stamp

東京 Tokyo

		**	●	FDC
C279	10 yen 多色 multicolored	50	40	400

切手データ Basic Information

版式：グラビア輪転版	printing plate : rotary photogravure plate
目打：櫛型13×13½	perforation : comb perf. 13×13½
印面寸法：40×22.5mm	printing area size : 40×22.5mm
発行枚数：1,700万枚	number issued : 17 million
シート構成：20枚（横5×縦4）	pane format : 20 subjects (5×4)
製版・目打組合せ：タイプ1E-1	plate and perforation configuration : type 1E-1

●スクリーン角度 Screen Angles

2版 Plate 2　緑 1度 green 1 degrees

1版 Plate 1　緑 89度 green 89 degrees

1）1版より2版の方が少ないと推測されるが、調査未了である。
1) Plate 2 is thought to be scarcer in comparison with plate 1, but this has not been confirmed.

C279 ●定常変種 Constant Flaws

P1-A-pos.8　　P1-A-pos.14　　P1-B-pos.16　　P1-B-pos.17

P1-C-pos.8　　P1-C-pos.16

P1-D-pos.6

位置 position	特徴 feature	**	●
P1-A-pos.8	"日"の左上に白抜き white outline above left of "日"	200	100
P1-A-pos.14	後部マスト上方に白点 white dot above rear mast		
P1-B-pos.16	"便"の上に大きな白点 large white dot above "便"		
P1-B-pos.17	"0"の右下に茶点 brown dot below right of "0"		
P1-C-pos.8	"1"の右上に白点 white dot above right of "1"		
P1-C-pos.16	印面左辺に茶点 brown dot at left edge of printing area		
P1-D-pos.6	"0"の右下方に大きな白点2ヵ所 two large white dots below right of "0"		

記念・特殊切手（額面「円位」時期、ローマ字なし）

C280 国際胸部医学・気管食道科学会議記念
International Congress of Thoracic Surgery and Broncho-Esophagology

発行日：1958（昭和33）.9.7　Date of Issue：1958 (Showa 33) .9.7

C280 聴診器 Stethoscope

●特印 Special Date Stamp
東京 Tokyo

	**	●	FDC
C280 10 yen 暗い青緑 prussian green	50	50	400

切手データ Basic Information	
版式：グラビア輪転版	printing plate : rotary photogravure plate
目打：櫛型13½	perforation : comb perf. 13½
印面寸法：33×22.5mm	printing area size : 33×22.5mm
発行枚数：1,500万枚	number issued : 15 million
シート構成：20枚（横5×縦4）	pane format : 20 subjects (5x4)
製版・目打組合せ：タイプ1C-2	plate and perforation configuration : type 1C-2

C280 ●定常変種 Constant Flaws

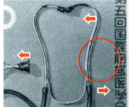
A-pos.1

A-pos.12

B-pos.2

B-pos.12

C-pos.2　C-pos.19　D-pos.5　D-pos.10

位置 position	特徴 feature	**	●
A-pos.1	白点、黒点、円弧状の白線、濃い点 white dot, black dot, curved white line and dark dot	200	100
A-pos.12	"会"に左に白点 white dot to left of "会"		
B-pos.2	聴診器の下に白抜け white spot below stethoscope		
B-pos.12	大きな薄い白点3ヵ所 three large thin white spots		
C-pos.2	聴診器の下に緑点 green dot below stethoscope		
C-pos.19	聴診器に裂けた傷 rip in stethoscope tube		
D-pos.5	聴診器近くに緑点と白点 green dot and white dot near stethoscope		
D-pos.10	聴診器の右に緑点 green dot to right of stethoscope		

C281 国際文通週間
International Letter Writing Week 1958

発行日：1958（昭和33）.10.5　Date of Issue：1958 (Showa 33) .10.5

C281 歌川広重画「東海道五十三次 京師（三条大橋）」 "The Fifty-three Stations of the Tokaido: Keishi (Sanjo Ohashi Bridge)" by Utagawa Hiroshige

●特印 Special Date Stamp
京都 Kyoto

	**	●	FDC
C281 24 yen 多色 multicolored	400	180	2,500

切手データ Basic Information	
版式：グラビア輪転版	printing plate : rotary photogravure plate
目打：櫛型13½	perforation : comb perf. 13½
印面寸法：35.5×25mm	printing area size : 35.5×25mm
発行枚数：800万枚	number issued : 8 million
シート構成：20枚（横5×縦4）	pane format : 20 subjects (5x4)
製版・目打組合せ：タイプ1D-1	plate and perforation configuration : type 1D-1

C281 ●定常変種 Constant Flaws

A-pos.6

C-pos.14

C-pos.20

D-pos.8

D-pos.13

D-pos.16

D-pos.18

D-pos.9

位置 position	特徴 feature	**	●
A-pos.6	"郵"の上に枠線欠け gap in frame line above "郵"	800	400
C-pos.14	右端の山に黒い塊 black mass on rightmost mountain		
C-pos.20	落款の右下方に青い煙 blue smoke below right of signature		
D-pos.8	落款の右に青点 blue dot to right of signature		
D-pos.9	落款の右に黒点、"東"の右に黒点 black dots to right of signature and to right of "東"		
D-pos.13	"R"の下方に黒点2個 two black dots below "R"		
D-pos.16	山頂に黒い汚れ black stain on the mountain top		
D-pos.18	"京師"の左下方に黒点群 black dots below left of "京師"		

記念・特殊切手（額面「円位」時期、ローマ字なし）

C282-283　第13回国体記念
13th National Athletic Meet

発行日：1958（昭和33）.10.19
Date of Issue : 1958 (Showa 33).10.19

C282 バドミントン Badminton
C283 重量挙げ Weightlifting

●特印 Special Date Stamp

富山 Toyama

	**	●	FDC
C282 5 yen こい赤 claret	50	40	
C283 5 yen 暗い青 gray blue	50	40	
C282-283 (2種連刷 pair)	120	160	400

切手データ Basic Information	
版式：凹版速刷版	printing plate : high-speed intaglio printing
目打：櫛型13½	perforation : comb perf. 13½
印面寸法：22.5×27mm	printing area size : 22.5×27mm
発行枚数：各1,500万枚	number issued : 15 million each
シート構成：20枚（横4×縦5）	pane format : 20 subjects (4x5)
製版・目打組合せ：タイプ1B-3	plate and perforation configuration : type 1B-3

C284　慶応義塾創立100年記念
Centenary of Keio University

発行日：1958（昭和33）.11.8
Date of Issue : 1958 (Showa 33).11.8

C284 塾舎と福沢諭吉像 Keio University and Fukuzawa Yukichi

●特印 Special Date Stamp

慶應義塾前 Keio University

	**	●	FDC
C284 10 yen 暗い赤 magenta	50	50	500

切手データ Basic Information	
版式：凹版速刷版	printing plate : high-speed intaglio printing
目打：櫛型13	perforation : comb perf. 13
印面寸法：22.5×33mm	printing area size : 22.5×33mm
発行枚数：1,500万枚	number issued : 15 million
シート構成：20枚（横4×縦5）	pane format : 20 subjects (4x5)
製版・目打組合せ：タイプ1B-3	plate and perforation configuration : type 1B-3

C285　国際児童福祉研究・社会事業会議記念
International Congress of Child Welfare Research and Social Work

発行日：1958（昭和33）.11.23
Date of Issue : 1958 (Showa 33).11.23

C285 地球と子供たち Glove and Playing Children

●特印 Special Date Stamp

東京 Tokyo

	**	●	FDC
C285 10 yen こい緑 deep green	50	50	350

切手データ Basic Information	
版式：グラビア輪転版	printing plate : rotary photogravure plate
目打：櫛型13	perforation : comb perf. 13
印面寸法：33.5×25mm	printing area size : 33.5×25mm
発行枚数：1,200万枚	number issued : 12 million
シート構成：20枚（横5×縦4）	pane format : 20 subjects (5x4)
製版・目打組合せ：タイプ1D-1	plate and perforation configuration : type 1D-1

C285 ●定常変種 Constant Flaws

P1/P2-A-pos.6　(P1)　P1/P2-A-pos.9　(P2)

P1-A-pos.9

P2-A-pos.9

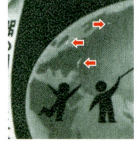

(P1)　P1/P2-A-pos.10　(P2)

P1/P2-B-pos.14

P1-B-pos.6

P2-B-pos.15　P1/P2-C-pos.5

P2-C-pos.4

記念・特殊切手（額面「円位」時期、ローマ字なし）

(P1)　P1/P2-D-pos.16　(P2)

P1-D-pos.2

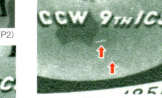
P2-D-pos.4

位置 position	特徴 feature	**	●
P1/P2-A-pos.6	オーストラリアの下に大きな緑点（タスマニア島）big green dot under Australia ("Tasmania")	500	200
P1/P2-A-pos.9	大西洋リタッチ retouch Atlantic Ocean		
P1-A-pos.9	"国際"の左に白の突起 white protrusion to left of "国際"		
P2-A-pos.9	ベーリング海に緑点 green dot in Bering Sea		
P1/P2-A-pos.10	子供の手の上に緑点（ハワイ島）green dot above child's hand ("Hawaii Island")		
P1/P2-B-pos.14	子供の手の右下方に緑点（にせハワイ島）green dot below right of child's hand ("fake Hawaii Island")		
P1-B-pos.6	第1コーナーに球状のシミ spherical stains in 1st corner		
P2-B-pos.15	"国"と地球の間に濃い緑点 dark green dot between "国" and earth		

P1/P2-C-pos.5	リボンの下に濃い緑点 dark green dot under ribbon		
P2-C-pos.4	カスピ海付近に大きな白抜き（カスピ海）、イギリスの右に白点、半島の付け根に白点 large white spot near Caspian Sea, white dots to right of England and at base of peninsula		
P1/P2-D-pos.16	子供の足の下に大きい白点 large white dot below child's foot		
P1-D-pos.2	"I"の左に緑点 green dot to left of "I"		
P2-D-pos.4	"9"の下方に白抜き斜線と白点 white diagonal line and white dot below "9"		

1）定常変種の現れ方から実用版上の同じ位置の窓口シートで2タイプが確認されているため、便宜的に1版と2版に分類した。原因は、印刷途中で版面を洗浄しクロムメッキをやり直して再印刷したのか、新たに実用版を製作したのかは断定が難しい。残存量は1版の方が少ないが、2つの版の時系列的な順番は不明である。「国際児童福祉記念切手の定常変種と版の研究」（青山宏、「郵趣研究」1992冬号）参照。

1) Although the number of plates was one, two types of panes, in which some of constant flaws are different, are known. These two panes are named "plate 1" and "plate 2" for convenience. It is difficult to clarify the reason why two types of panes exist; one possibility is that plate was re-chrome-plated during printing; another other possibility is that a new plate was made during the printing. The existing proportion of "plate 1" is smaller than that of "plate 2". (Hiroshi Aoyama, "Study of constant flaws and plates of 'International Congress of Child Welfare Research and Social Work'", The Philatelic Study, Winter, 1992)

C286 世界人権宣言10年記念
10th Anniversary of Human Rights Declaration

発行日：1958（昭和33）.12.10　Date of Issue：1958 (Showa 33).12.10

C286 人権をあらわす赤い炎 Symbol of Human Rights

●特印 Special Date Stamp

東京 Tokyo

	**	●	FDC
C286 10 yen 多色 multicolored	60	50	350

切手データ Basic Information

版式：グラビア輪転版	printing plate : rotary photogravure plate
目打型：櫛型13	perforation : comb perf. 13
印面寸法：25×35.5mm	printing area size : 25×35.5mm
発行枚数：1,500万枚	number issued : 15 million
シート構成：20枚（横4×縦5）	pane format : 20 subjects (4×5)
製版・目打組合せ：タイプ1D-4	plate and perforation configuration : type 1D-4

C286 ●定常変種 Constant Flaws

A-pos.3　B-pos.12　B-pos.13　C-pos.1

C-pos.5　C-pos.9　C-pos.13
　　　　　　　　　　　D-pos.5

位置 position	特徴 feature	**	●
A-pos.3	"宣言"の上方に白抜き群 white spots above "宣言"	200	100
B-pos.12	第2コーナーに大きなリタッチ群 large retouches at NE corner		
B-pos.13	炎の右上に白点2ヵ所 two white dots to upper right of flame		
C-pos.1	炎の右に濃い帯状の筋と白点群 dark streaks and white dots to right of flame		
C-pos.5	"日本郵便"の下に斜線状のリタッチ retouched line under "日本郵便"		
C-pos.9	第2コーナーに白点2個 two white dots in second corner		
C-pos.13	第2コーナーに大きなリタッチ群 large retouches at NE corner		
D-pos.5	上部マージンに濃い点と炎の右下に白点 dark dot in top margin and white dot below right of flame		

記念・特殊切手（額面「円位」時期、ローマ字なし）

C287 児島湾締切堤防竣工記念　発行日：1959（昭和34）.2.1
Completion of Kojima Bay Cutoff Embankment　Date of Issue：1959 (Showa 34) .2.1

C287 干拓地の地図とトラクター
Tractor and Map of Kojima Bay

●特印
Special Date Stamp

岡山 Okayama

	**	●	FDC
C287 10 yen 茶紫・茶 claret and brown …… 50	50	300	

切手データ Basic Information	
版式：グラビア輪転版	printing plate : rotary photogravure plate
目打：櫛型12½	perforation : comb perf. 12½
印面寸法：24×27mm	printing area size : 24×27mm
発行枚数：1,200万枚	number issued : 12 million
シート構成：20枚（横4×縦5）	pane format : 20 subjects (4×5)
製版・目打組合せ：タイプ1B-3	plate and perforation configuration : type 1B-3

C287 ●定常変種 Constant Flaws

 A-pos.19
 B-pos.4
 B-pos.19
 C-pos.3
 C-pos.15
 D-pos.13
 D-pos.16

位置 position	特徴 feature	**	●
A-pos.19	"念"の上方に斜め紫破線 claret dashed line above "念"	200	100
B-pos.4	煙突の右に紫点 claret dot to right of exhaust pipe		
B-pos.19	煙突の上方に小紫点、"切"の上に小紫点 small claret dot above exhaust pipe, small claret dot above "切"		
C-pos.3	地図の右に白点 white dot to right of map		
C-pos.15	地図の右に紫線 claret line to right of map		
D-pos.13	"9"の上方に茶点 claret dot above "9"		
D-pos.16	紫の連点と紫の小点群 continuous line of claret dots and group of small claret dots		
D-pos.18	"郵"の上に紫点 claret dot above "郵"		

D-pos.18

C288 アジア文化会議記念　発行日：1959（昭和34）.3.27
Asian Cultural Congress on 2,500th Death Anniversary of Buddha　Date of Issue：1959 (Showa 34) .3.27

C288 東南アジア地図
Map of Southeast Asia

●特印
Special Date Stamp

東京 Tokyo

	**	●	FDC
C288 10 yen こい赤 deep carmine	50	50	300

切手データ Basic Information	
版式：グラビア輪転版	printing plate : rotary photogravure plate
目打：櫛型13½	perforation : comb perf. 13½
印面寸法：22.5×27mm	printing area size : 22.5×27mm
発行枚数：1,200万枚	number issued : 12 million
シート構成：20枚（横4×縦5）	pane format : 20 subjects (4×5)
製版・目打組合せ：タイプ1B-3	plate and perforation configuration : type 1B-3

C288 ●定常変種 Constant Flaws

 A-pos.4
 A-pos.19
 B-pos.6
 B-pos.8
 C-pos.16
 C-pos.19
 D-pos.4
 D-pos.8

位置 position	特徴 feature	**	●
A-pos.4	"9"の上方に赤点 carmine dot above "9"	200	100
A-pos.19	"本"の下方に白点 white dot below "本"		
B-pos.6	"59"の上方に赤点 carmine dot above "59"		
B-pos.8	パプアニューギニアの上方に2本の曲線 two curved lines above Papua New Guinea		
C-pos.16	北朝鮮の上部に白点 white dot above North Korea		
C-pos.19	"5"の上方に赤点3個 three carmine dots above "5"		
D-pos.4	マレー半島の上部に赤点 carmine dot on upper part of Malay Peninsula		
D-pos.8	ミャンマー沖に大赤点と小赤点 large and small carmine dots off the coast of Myanmar		

記念・特殊切手（額面「円位」時期、ローマ字なし）

C289-292 皇太子（明仁）御成婚記念　発行日：1959（昭和34）.4.10
Wedding of Crown Prince Akihito and Princess Michiko　Date of Issue：1959 (Showa 34) .4.10

C289　檜扇 Ceremonial Fan

C291

C290　皇太子夫妻の肖像 Prince Akihito and Princess Michiko

C292

●特印 Special Date Stamp

東京 Tokyo

	**	●	FDC
C289 5 yen 紫・赤紫 magenta and violet	50	40	200
C290 10 yen 赤茶・暗い赤茶 red brown and dull purple	100	50	200
C291 20 yen だいだい黄・うす茶 orange brown and brown	150	50	500
C292 30 yen 緑・暗い緑 yellow green and dark green	600	70	500
C289-292 (4種 set of 4 values)	900	210	1,400

切手データ Basic Information

版式：C289、C291＝グラビア輪転版、C290、C292＝グラビア輪転版・凹版速刷版	printing plate：C289,C291=rotary photogravure plate, C290,C292=rotary photogravure plate and high-speed intaglio printing
目打：櫛型13½	perforation：comb perf. 13½
印面寸法：33×22.5mm	printing area size：33×22.5mm
発行枚数：C289、C290＝各2,500万枚、C291、C292＝各1,500万枚	number issued：C289,C290=25 million each, C291,C292=15 million each
シート構成：各20枚（横5×縦4）	pane format：20 subjects (5x4) each
製版・目打組合せ：C289,C291＝タイプ1C-1、C290＝タイプ1B-1、C292＝タイプ1B-1、1C-1	plate and perforation configuration：C289,C291=type 1C-1、C290=type 1B-1、C292=type 1B-1 and 1C-1

●定常変種 Constant Flaws

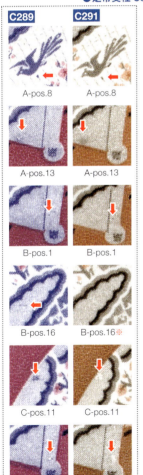

※＝版欠点なし normal

C289

位置 position	特徴 feature	**	●
A-pos.8	紫点 magenta dot	200	100
A-pos.13	広いリタッチ wide retouch		
B-pos.1	紫の小点群 small magenta dots		
B-pos.16	紫線 magenta line		
C-pos.11	リタッチ retouch		
C-pos.14	白抜け・リタッチ white spot and retouch		
C-pos.16	紫線 magenta line		
D-pos.5	紫点 magenta dot		
D-pos.7	紫点 magenta dot		
D-pos.11	白抜け・リタッチ white spot and retouch		
D-pos.15	紫点 magenta dot		

C291

位置 position	特徴 feature	**	●
A-pos.8	茶点 brown dot	300	150
A-pos.13	広いリタッチ wide retouch		
B-pos.1	茶点（小点群なし） brown dot (no small dots)		
C-pos.11	白抜け（リタッチなし） white spot (no retouch)		
C-pos.14	白抜け（リタッチなし） white spot (no retouch)		
D-pos.11	白抜け・リタッチ white spot and retouch		
D-pos.15	茶点 brown dot		

1) C289とC291は同一の原乾板から実用版が製作されたので同一の定常変種が存在するが、一部のポジションの定常変種は製版の過程で消滅している。

1) Since the plates of C289 and C291 were made from same multi-positive, these two stamps share some constant flaws. However, some constant flaws disappeared during the plate making process.

記念・特殊切手（額面「円位」時期、ローマ字なし）

C290 ●定常変種 Constant Flaws

A-pos.5　　　B-pos.5　　　　　B-pos.7

C-pos.3　　C-pos.15　　C-pos.16　　D-pos.4

位置 position	特徴 feature	**	●
A-pos.5	"日"の左下方に大きな白点 large white dot at lower left of "日"	400	150
B-pos.5	白点 white dot		
B-pos.7	白点2ヵ所 two white dots		
C-pos.3	白点 white dot		
C-pos.15	髪の右上に茶点 brown dot at upper right of hair		
C-pos.16	茶点と白抜き brown dot and white spot		
D-pos.4	"郵"の右上に茶点 brown dot at upper right of "郵"		

1) グラビアと凹版の印刷ズレによっては、定常変種の位置が上下左右に異なる。
1) Depending on the misalignment between the photogravure and intaglio printings, the position of a constant flaw may differ up, down, left, or right.

C292
●定常変種 Constant Flaws

A-pos.2　　A-pos.3　　A-pos.11　　A-pos.15

(type1C-1)　A-pos.15　(type1B-1)

C-pos.15　　C-pos.17　　D-pos.1　　D-pos.2

位置 position	特徴 feature	**	●
A-pos.2	小緑点 small green dot	1,200	200
A-pos.3	緑点 green dot		
A-pos.11	太い緑線 thick green line		
A-pos.15	白点 white dot (type1B-1)		
A-pos.15	緑点 green dot (type1C-1)		
B-pos.5	白点 white dot		
I-C-pos.8	白点 white got		
I-C-pos.15	白抜け・リタッチ white spot and retouch		
I-C-pos.17	白点群 white dots		
D-pos.1	白点 white dot		
D-pos.2	大きな白抜け large white spot		

1) C292には製版・目打組合せにタイプ1B-1とタイプ1C-1の2種があり、A-pos.15には別々の定常変種がある。
2) シートCには製版・目打組合せにタイプ1B-1とタイプ1C-1の2種があるが、共通の変種はない。
There are two types of plate and perforation configurations, types 1B-1 and 1C-1, of C292. The two types have different constant flaws at A-pos. 15.
2) There are two types of plate and perforation configurations, 1B-1 and 1C-1, for Pane C but they do not share any common flaws.

●目打の抜け方 Position of perforations through selvages

右抜け right selvage only

左右抜け both side selvages

1) C292は、当初は製版・目打組合せにタイプ1C-1が採用されたが、すぐにタイプ1B-1に変更されたため、タイプ1C-1の左右抜けは少ない。
1) The plate and perforation configuration for C292 was type 1C-1 at first, but it was changed to type 1B-1 in the early stages of printing. Therefore, the perforation through both right and left selvages of 1C-1 is scarce.

記念・特殊切手（額面「円位」時期、ローマ字なし）

C293　皇太子(明仁)御成婚記念 小型シート　発行日：1959（昭和34）.4.20
Wedding of Crown Prince Akihito and Princess Michiko, Souvenir Sheet　Date of Issue：1959 (Showa 34) .4.20

C293 小型シート
Souvenir Sheet

タトウ表紙
Front of Folder

	**	●	FDC
C293 売価20円 selling price 20 yen	1,200	1,200	3,500
C293a 5 yen 目打・糊なし without perforation and gum	420	360	
C293b 10 yen 目打・糊なし without perforation and gum	540	360	

切手データ Basic Information	
版式：グラビア輪転版・凹版速刷版	printing plate：rotary photogravure plate and high-speed intaglio printing
目打・糊なし	perforation：without perforation and gum
シート寸法：128×88mm	sheet size：128×88mm
発行枚数：200万枚	number issued：2 million

C293b ● 定常変種 Constant Flaws

C293b-2

C293b-7　C293b-8

位置 position	特徴 feature	**	●
C293b-1	定常変種なし not constant flaw	1,200	1,200
C293b-2	"日"の左に白点と茶点、茶点2ヵ所、襟に茶点 white dot and brown dot to left of "日", two brown dots, brown dot on collar		
C293b-3	茶点、白点 brown dot and white dot		
C293b-4	白抜きの突起、茶点 white protrusion and brown dot		
C293b-5	茶点2ヵ所 two brown dots		
C293b-6	白点、白抜け、茶リタッチ white dot, white spot and brown retouch		
C293b-7	花弁に茶点 brown dot on petals		
C293b-8	花弁に茶点 brown dot on petals		

1）グラビア輪転機で3色刷り後に凹版速刷機により1色刷りされているが、グラビア輪転機での製版上の制約から、2×4の8面掛けまたは2×6の12面掛けで印刷されたものと推定される。定常変種により8面分まで確認できている。

1) After three colors were printed by rotary photogravure, one color was printed by high-speed intaglio printing. The plate configuration is thought to be 2×4 (8 panes) or 2×5 (10 panes) due to limitations of the rotary photogravure printing. So far, eight panes have been identified by the constant flaws.

C294　切手趣味週間　発行日：1959（昭和34）.5.20
Philatelic Week 1959　Date of Issue：1959 (Showa 34) .5.20

C294
細田栄之画
「浮世源氏八景」
"Eight Views of Genji in the Floating World"
by Hosoda Eishi

● 特印
Special Date Stamp

東京 Tokyo

	**	●	FDC
C294 10 yen 多色 multicolored	240	240	900

切手データ Basic Information	
版式：グラビア輪転版	printing plate：rotary photogravure plate
目打：櫛型13½	perforation：comb perf. 13½
印面寸法：33×48mm	printing area size：33×48mm
発行枚数：1,500万枚	number issued：15 million
シート構成：10枚（横5×縦2）	pane format：20 subjects (5x2)
製版・目打組合せ：タイプ1C-2	plate and perforation configuration：type 1C-2

次ページにつづく Continued to next page ↗

41

記念・特殊切手（額面「円位」時期、ローマ字なし）

C294
●定常変種
Constant Flaws

位置 position	特徴 feature	**	●
A-pos.5	顔の左方に黒点、髪の右方に白点 black dot to left of face, white dot to right of hair	600	600
B-pos.4	文台の近くに白抜き white spot near writing desk		
C-pos.3	髪の左方に緑点 green dot to left of hair		
D-pos.8	マージンに黄色の太線 thick yellow line in margin		

C295 メートル法完全実施記念
Adoption of Metric System
発行日：1959（昭和34）.6.5　　Date of Issue：1959 (Showa 34) .6.5

C295 ものさし・はかり・ます
Tape Measure, Scales and Graduated Beaker

●特印 Special Date Stamp

東京 Tokyo

	**	●	FDC
C295 10 yen にぶ青・茶・黒 light blue, brown and black	50	50	300

切手データ Basic Information	
版式：グラビア輪転版	printing plate : rotary photogravure plate
目打：櫛型13½	perforation : comb perf. 13½
印面寸法：22.5×27mm	printing area size : 22.5×27mm
発行枚数：1,200万枚	number issued : 12 million
シート構成：20枚（横4×縦5）	pane format : 20 subjects (4x5)
製版・目打組合せ：タイプ1B-3	plate and perforation configuration : type 1B-3

C295 ●定常変種 Constant Flaws

A-pos.2　A-pos.12　A-pos.20　B-pos.16

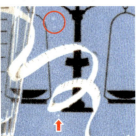

B-pos.20　　C-pos.1　C-pos.15　D-pos.14

D-pos.15　D-pos.17

位置 position	特徴 feature	**	●
A-pos.2	"1"の左に白線 white line to left of "1"	200	100
A-pos.12	白点と太い白斜線 white dots and thick white diagonal line		
A-pos.20	"0"の上に白傷 white scratch on "0"		
B-pos.16	"全"の上方に濃い青染、"施"の左に白抜け dark blue stain above "全", white spot to left of "施"		
B-pos.20	巻尺の上方に白点3個、下方に白点 three white dots above tape measure, white dot below tape measure		
C-pos.1	"全"の上方に白点 white dot above "全"		
C-pos.15	"0"の下方に白点2個 two dots below "0"		
D-pos.14	"0"の下方と天秤の右に白点 white dots below "0" and to right of balance		
D-pos.15	"0"の中に茶線 brown line inside "0"		
D-pos.17	"日"の左と"本"の下方に白点 white dots to left of "日" and below "本"		

C296 赤十字思想誕生100年記念
Centenary of Idea of Red Cross
発行日：1959（昭和34）.6.24　　Date of Issue：1959 (Showa 34) .6.24

C296 看護師の活動 Nurses Carrying Stretcher

●特印 Special Date Stamp

東京 Tokyo

	**	●	FDC
C296 10 yen 暗い緑・赤 olive green and red	50	50	300

切手データ Basic Information	
版式：グラビア輪転版	printing plate : rotary photogravure plate
目打：櫛型13½	perforation : comb perf. 13½
印面寸法：33×22.5mm	printing area size : 33×22.5mm
発行枚数：1,200万枚	number issued : 12 million
シート構成：20枚（横5×縦4）	pane format : 20 subjects (5x4)
製版・目打組合せ：タイプ1C-2	plate and perforation configuration : type 1C-2

記念・特殊切手（額面「円位」時期、ローマ字なし）

C296 ●定常変種 Constant Flaws

A-pos.6　B-pos.1　B-pos.6　B-pos.18

C-pos.13　D-pos.12　D-pos.17　D-pos.18

位置 position	特徴 feature	**	●
A-pos.6	"便"の右と顔の左に緑点 green dots to right of "便" and to left of face	200	100
B-pos.1	"本"の左下に緑点2個 two green dots to lower left of "本"		
B-pos.6	赤十字に白点 white dot on red cross		
B-pos.18	"記"の上に小白点、"0"に白突起 small white dot above "記", white protrusion on "0"		
C-pos.13	"郵"の左下に緑点 green dot to lower left of "郵"		
D-pos.12	"9"の下方に緑点 green dot below "9"		
D-pos.17	看護師の背中の右方に緑点 green dot to right of nurse's back		
D-pos.18	"生"の上方に緑点 green dot above "生"		

C297 自然公園の日制定記念 / Establishment of Natural Park Day

発行日：1959（昭和34）.7.21　Date of Issue : 1959 (Showa 34) .7.21

C297 本栖湖と富士山 Lake Motosu and Mt. Fuji

●特印 Special Date Stamp
日光 Nikko

	**	●	FDC
C297 10 yen 緑・赤・青 green, sepia and blue	80	50	350

切手データ Basic Information	
版式：ザンメル凹版	printing plate : sammeldruck intaglio printing
目打：櫛型13½	perforation : comb perf. 13½
印面寸法：33×22.5mm	printing area size : 33×22.5mm
発行枚数：1,500万枚	number issued : 15 million
シート構成：20枚（横5×縦4）	pane format : 20 subjects (5×4)
製版・目打組合せ：タイプ1B-2	plate and perforation configuration : type 1B-2

1) 初めてのザンメル凹版多色刷の切手である。使われた印刷機は『新版「切手と印刷」』では凹版速刷機を改造したと記されているが、後に小型凹版輪転機をザンメル印刷できるように改造した機械が凹版印刷切手のすべてに使用された。

1) This was the first stamp that was printed by the sammeldruck intaglio printing process in Japan. According to "Stamps and Printing, New Edition", the press for this sammeldruck intaglio printing was a modification of the high-speed intaglio press. Later, all the intaglio printing stamps were printed by a sammeldruck intaglio press modified from a small rotary intaglio press.

C298 名古屋開府350年記念 / 350th Anniversary of Nagoya

発行日：1959（昭和34）.10.1　Date of Issue : 1959 (Showa 34) .10.1

C298 金の鯱と名古屋市街 Golden Shachi (roof finial) and Nagoya city scape

●特印 Special Date Stamp
名古屋 Nagoya

	**	●	FDC
C298 10 yen 青・黒・金 bright blue, black and gold	50	50	300

切手データ Basic Information	
版式：グラビア輪転版	printing plate : rotary photogravure plate
目打：櫛型13½	perforation : comb perf. 13½
印面寸法：27×22.5mm	printing area size : 27×22.5mm
発行枚数：1,200万枚	number issued : 12 million
シート構成：20枚（横5×縦4）	pane format : 20 subjects (5×4)
製版・目打組合せ：タイプ1C-2	plate and perforation configuration : type 1C-2

C298 ●定常変種 Constant Flaws

A-pos.19　B-pos.1　B-pos.5

B-pos.15　C-pos.7　D-pos.4

位置 position	特徴 feature	**	●
A-pos.19	尾ひれの左に白点 white dot to left of tail fin	200	100
B-pos.1	タワーの右に白点 white dot to right of tower		
B-pos.5	"名"の左上方に黒煙、第3マージンに黒点 black smoke to upper left of "名", black dot in SE margin		
B-pos.15	尾ひれの左方に白点 white dot to left of tail fin		
C-pos.7	"年"の上方に白い曲線 white curve upper "年"		
D-pos.4	"1959"の左上方に青点2ヵ所、第3マージンに太い黒線 two blue dots to upper left of "1959", thick black line in SE margin		

記念・特殊切手（額面「円位」時期、ローマ字なし）

C299 国際文通週間
International Letter Writing Week 1959

発行日：1959（昭和34）.10.4
Date of Issue : 1959 (Showa 34) .10.4

C299 歌川広重画「東海道五十三次 桑名」
"The Fifty-three Stations of the Tokaido: Kuwana" by Utagawa Hiroshige

●特印 Special Date Stamp

東京 Tokyo

	**	●	FDC
C299 10 yen 多色 multicolored	1,800	300	3,000

切手データ Basic Information	
版式：グラビア輪転版	printing plate : rotary photogravure plate
目打：櫛型13	perforation : comb perf. 13
印面寸法：35.5×25mm	printing area size : 35.5×25mm
発行枚数：800万枚	number issued : 8 million
シート構成：20枚（横5×縦4）	pane format : 20 subjects (5x4)
製版・目打組合せ：タイプ1D-2	plate and perforation configuration : type 1D-2

C299 ●定常変種 Constant Flaws

A-pos.1　　A-pos.17　　C-pos.2　　C-pos.14

C-pos.16　　D-pos.3　　D-pos.7

位置 position	特徴 feature	**	●
A-pos.1	題名の下にU形の黒曲線 black curved line below title	4,000	600
A-pos.17	"3"の左上方に青点 blue dot to upper left of "3"		
C-pos.2	帆船に線切れ、帆船の左上方に黒点 broken line on sailboat, black dot to upper left of sailboat		
C-pos.14	落款の右に黒汚れ black stain to right of seal		
C-pos.16	"郵"の上方に黒点 black dot above "郵"		
D-pos.3	帆の上に黒点 black dot above sail		
D-pos.7	題名の右下に黒点 black dot to lower right of title		

C300 第15回国際航空運送協会総会記念
15th General Meeting of International Air Transport Association

発行日：1959（昭和34）.10.12
Date of Issue : 1959 (Showa 34) .10.12

C300 タンチョウとIATAマーク
Japanese Crane and IATA Symbol

●特印 Special Date Stamp

東京 Tokyo

	**	●	FDC
C300 10 yen 青 blue	50	50	400

切手データ Basic Information	
版式：凹版速刷版	printing plate : high-speed intaglio printing
目打：櫛型13½	perforation : comb perf. 13½
印面寸法：27×23mm	printing area size : 27×23mm
発行枚数：1,000万枚	number issued : 10 million
シート構成：20枚（横5×縦4）	pane format : 20 subjects (5x4)
製版・目打組合せ：タイプ1C-2	plate and perforation configuration : type 1C-2

C300 ●定常変種 Constant Flaws

位置 position	特徴 feature	**	●
A-pos.10	"T"の右に縁欠け nick in edge of design to right of "T"	200	100

C301-302 第14回国体記念
14th National Athletic Meet

発行日：1959（昭和34）.10.25
Date of Issue : 1959 (Showa 34) .10.25

C301 ハンマー投げ Hammer Throw　　C302 フェンシング Fencing

●特印 Special Date Stamp

東京 Tokyo

	**	●	FDC
C301 5 yen 暗い青緑 gray blue	60	40	
C302 5 yen 暗い茶 olive bister	60	40	
C301-302 (2種連刷 pair)	150	160	450

切手データ Basic Information	
版式：ザンメル凹版	printing plate : sammeldruck intaglio printing
目打：櫛型13½	perforation : comb perf. 13½
印面寸法：27×22.5mm	printing area size : 27×22.5mm
発行枚数：各1,000万枚	number issued : 10 million each
シート構成：20枚（横5×縦4）	pane format : 20 subjects (5x4)
製版・目打組合せ：タイプ1B-3	plate and perforation configuration : type 1B-3

記念・特殊切手（額面「円位」時期、ローマ字なし）

C303 松陰100年祭 PTA大会記念　発行日：1959（昭和34）.10.27
Death Centenary of Yoshida Shoin and Convention of Parent-Teacher Associations
Date of Issue : 1959 (Showa 34).10.27

C303 吉田松陰像とPTAマーク
Yoshida Shoin and PTA Symbol

●特印 Special Date Stamp

萩 Hagi

	**	○	FDC
C303 10 yen 茶 brown	50	50	400

切手データ Basic Information	
版式：グラビア輪転版	printing plate : rotary photogravure plate
目打：櫛型13½	perforation : comb perf. 13½
印面寸法：27×22.5mm	printing area size : 27×22.5mm
発行枚数：1,000万枚	number issued : 10 million
シート構成：20枚（横5×縦4）	pane format : 20 subjects (5x4)
製版・目打組合せ：タイプ1C-2	plate and perforation configuration : type 1C-2

C303 ●定常変種 Constant Flaws

 A-pos.4
 A-pos.15
 B-pos14
 B-pos.18
 B-pos.20
 C-pos.16
 C-pos.19
 D-pos.1
 D-pos.2

位置 position	特徴 feature	**	○
A-pos.4	"19"の下方に茶点 brown dot below "19"	200	100
A-pos.15	マーク近くに茶点2ヵ所、本の縁に曲線 two brown dots near symbol, curved lines on edge of book		
B-pos.14	本の左ページに茶点2個、本の右ページに茶点2ヵ所 two brown dots on left page of book, two brown dots on right page of book		
B-pos.18	マージンに茶点 brown dot in margin		
B-pos.20	マークの上方に茶点2個 two brown dots above symbol		
C-pos.16	膝の右上に茶点、"便"の左右に白点 brown dot to upper right of knee, white dot below left of "便"		
C-pos.19	本の縁に長い茶線 long brown line on edge of book		
D-pos.1	マークの左下に茶曲線、"9"の右上方に茶点 brown curve below left of mark, brown dot to upper right of "9"		
D-pos.2	顔の近くに茶点3ヵ所、"便"の下方に茶の染み three brown dots near face, brown stain below "便"		

C304 第15回ガット（GATT）東京総会記念　発行日：1959（昭和34）.11.2
15th GATT (General Agreement on Tariffs and Trade) Tokyo Congress
Date of Issue : 1959 (Showa 34).11.2

C304 地球と記念文字 Globes and Commemorative Characters

●特印 Special Date Stamp

東京 Tokyo

	**	○	FDC
C304 10 yen 赤茶 brown red	50	50	300

切手データ Basic Information	
版式：グラビア輪転版	printing plate : rotary photogravure plate
目打：櫛型13½	perforation : comb perf. 13½
印面寸法：36×25.5mm	printing area size : 36×25.5mm
発行枚数：1,000万枚	number issued : 10 million
シート構成：20枚（横5×縦4）	pane format : 20 subjects (5x4)
製版・目打組合せ：タイプ1C-2	plate and perforation configuration : type 1C-2

C304 ●定常変種 Constant Flaws

 A-pos.7
 A-pos.19
 B-pos.11　B-pos.15
 C-pos.4
 D-pos.1/6
 D-pos.18
 D-pos.20

位置 position	特徴 feature	**	○
A-pos.7	印面右上部に白点3ヵ所 three white dots in upper right part of printing area	200	100
A-pos.19	印面3ヵ所に白点 three white dots in printing area		
B-pos.11	上の地球に濃い茶点線 dark brown dotted line on top hemisphere		
B-pos.15	"C"の中に小さい白点 small white dot inside "C"		
C-pos.4	アフリカ大陸の上方に濃い茶線 dark brown line above African continent		
D-pos.1/6	北米大陸の右方に茶点2個 two brown dots to right of North American continent		
D-pos.18	下方のガッターに島 Island in lower gutter		
D-pos.20	アラスカの下方に茶点 brown dot below Alaska		

45

記念・特殊切手（額面「円位」時期、ローマ字なし）

C305 尾崎記念会館竣工記念 / Completion of Ozaki Memorial Hall
発行日：1960（昭和35）.2.25　　Date of Issue：1960 (Showa 35) .2.25

C305 尾崎行雄と記念館時計塔
Ozaki Yukio and Clock Tower

●特印 Special Date Stamp

東京 Tokyo

C305 10 yen セピア・くすみ赤 red blown and dark brown ……………………………………… 50　50　270
　　　　　　　　　　　　　　　　　　　　　　　　　　　　　　＊＊　●　FDC

C305 ●定常変種 Constant Flaws

A-pos.13　　B-pos.8　　B-pos.19

C-pos.1　　C-pos.14　　D-pos.20

切手データ Basic Information	
版式：グラビア輪転版	printing plate : rotary photogravure plate
目打：櫛型13½	perforation : comb perf. 13½
印面寸法：22.5×27mm	printing area size : 22.5×27mm
発行枚数：800万枚	number issued : 8 million
シート構成：20枚（横4×縦5）	pane format : 20 subjects (4×5)
製版・目打組合せ：タイプ1B-3	plate and perforation configuration : type 1B-3

位置 position	特徴 feature	**	●
A-pos.13	時計の左に白点 white dot to left of clock	200	100
B-pos.8	襟の右に茶点 brown dot to right of collar		
B-pos.19	マージンに小茶点 small brown dot in margin		
C-pos.1	"10"の上方に濃い茶点 dark brown dot above "10"		
C-pos.14	時計台の上に白点 white spot above clock tower		
D-pos.20	"日"の下に茶点 brown dot under "日"		

C306 奈良遷都1250年記念 / 1,250th Anniversary of Relocation of Capital to Nara
発行日：1960（昭和35）.3.10　　Date of Issue：1960 (Showa 35) .3.10

C306 正倉院「鹿草木夾纈屏風」（板締め染め）の鹿
"Kyōkechi dyed folding screen panel with deer, tree, and plant design" in the Shosoin treasury

●特印 Special Date Stamp

東京 Tokyo

C306 10 yen 灰味オリーブ olive gray ………… 50　50　250
　　　　　　　　　　　　　　　　　　　　　＊＊　●　FDC

切手データ Basic Information	
版式：グラビア輪転版	printing plate : rotary photogravure plate
目打：櫛型13½	perforation : comb perf. 13½
印面寸法：22.5×27mm	printing area size : 32.5×27mm
発行枚数：1,000万枚	number issued : 10 million
シート構成：20枚（横4×縦5）	pane format : 20 subjects (4×5)
製版・目打組合せ：タイプ1B-3	plate and perforation configuration : type 1B-3

C306 ●定常変種 Constant Flaws

A-pos.9　　B-pos.3　　B-pos.4　　B-pos.13

B-pos.19　　C-pos.5　　D-pos.4　　D-pos.20

位置 position	特徴 feature	**	●
A-pos.9	"良"の下に大きな白点 large white dot under "良"	200	100
B-pos.3	"1"と"0"の間に白点 white dot between "1" and "0"		
B-pos.4	"日"の左下に白点 white dot to lower left of "日"		
B-pos.13	"記"に斜線 diagonal line in "記"		
B-pos.19	"年"と"記"の間に斜線 diagonal line between "年" and "記"		
C-pos.5	"0"の上に大きな白抜け large white spot above "0"		
D-pos.4	印面右辺欠け nick in printing area at right		
D-pos.20	円盤状の太い線 thick disk shape in background		

記念・特殊切手（額面「円位」時期、ローマ字なし）

C307-309 日本三景
Three Famous Sights of Japan

発行日：C307=1960（昭和35）.3.15、C308=1960.7.15、C309=1960.11.15
Date of Issue : C307=1960 (Showa 35) .3.15, C308=1960.7.15, C309=1960.11.15

C307 松島の五大堂
Godaido Temple on Matsushima

C308 天橋立 Amanohashidate ("Bridge to Heaven")

C309 宮島の厳島神社 Itsukushima Shrine on Miyajima

●風景印 Pictorial Postmarks

P307 松島 Matsushima　　P308 天橋立 Amanohashidate　　P309 宮島 Miyajima

	**	● FDC
C307 10 yen 赤茶・青緑 maroon and blue green	200	120　650
C308 10 yen 暗い緑・青 green and light blue	300	120　700
C309 10 yen 緑・こい紫 violet black and blue green	300	120　750
P307-309（3種 set of 3 values）	800	360

版式：ザンメル凹版	printing plate : sammeldruck intaglio printing
目打：櫛型13½	perforation : comb perf. 13½
印面寸法：33×22.5mm	printing area size : 33×22.5mm
発行枚数：各800万枚	number issued : 8 million each
シート構成：20枚（横5×縦4）	pane format : 20 subjects (5x4)
製版・目打組合せ：タイプ1B-2	plate and perforation configuration : type 1B-2

C310 切手趣味週間
Philatelic Week 1960

発行日：1960（昭和35）.4.20
Date of Issue : 1960 (Showa 35) .4.20

C310「三十六歌仙絵巻 伊勢」
"Thirty-six Immortal Poets: Ise (poet)"

●特印 Special Date Stamp

大阪 Osaka

	**	● FDC
C310 10 yen 多色 multicolored	400	400　1,800

版式：グラビア輪転版	printing plate : rotary photogravure plate
目打：櫛型13½	perforation : comb perf. 13½
印面寸法：48×33mm	printing area size : 48×33mm
発行枚数：1,000万枚	number issued : 10 million
シート構成：10枚（横2×縦5）	pane format : 10 subjects (2x5)
製版・目打組合せ：タイプ1C-4	plate and perforation configuration : type 1C-4

C310 ●定常変種 Constant Flaws

A-pos.8　　B-pos.5　　C-pos.1　　C-pos.4　　C-pos.10

D-pos.5

位置 position	特徴 feature	**	●
A-pos.8	装束に灰色点 gray dot on costume	1,000	1,000
B-pos.5	装束の上に黒点 black dot above costume		
C-pos.1	"0"の中に灰色点 gray dot inside "0"		
C-pos.4	"0"の右上に灰色点 gray dot above right of "0"		
C-pos.10	"1"の下に灰色点 gray dot under "1"		
D-pos.5	小黒点 small black dot		
D-pos.6	頬に赤点 red dot on cheek		

D-pos.6

記念・特殊切手（額面「円位」時期、ローマ字なし）

C311-313 日米修好通商100年記念
Centenary of Treaty of Amity and Commerce Between U.S. and Japan

発行日：C311-312＝1960（昭和35）.5.17、C313＝1960（昭和35）.9.27
Date of Issue：C311-312＝1960 (Showa 35).5.17, C313＝1960 (Showa 35).9.27

C311 咸臨丸 Kanrin-Maru

C312 大統領の引見 Audience with President of U.S.

●特印 Special Date Stamp

東京 Tokyo

	**	●	FDC
C311 30 yen 暗い茶・青緑 blue green and brown	80	60	
C312 10 yen 紅・暗い青 carmine and indigo	300	100	
C311-312（2種連刷 pair）	380	160	850
C313 小型シート（売価40円）souvenir sheet (selling price 40 yen)	5,500	5,500	7,000

C313 小型シート souvenir sheet

切手データ Basic Information

版式：ザンメル凹版	printing plate : sammeldruck intaglio printing
目打：C311、C312＝櫛型13½ C313＝全型13½	perforation : C311、C312＝comb perf. 13½ C313＝harrow perf. 13½
印面寸法（C311、C312）：27×22.5mm	printing area size (C311、C312) : 27×22.5mm
シート寸法（C313）：120×76.5mm	sheet size (C313) : 120×76.5mm
発行枚数：C311＝1,000万枚、C312＝700万枚、C313＝50万枚	number issued : C311＝10 million, C312＝7 million, C313＝500,000
シート構成（C311、C312）：各20枚（横5×縦4）	pane format (C311、C312) : 20 subjects (5x4) each
製版・目打組合せ（C311、C312）：タイプ1B-3	plate and perforation configuration (C311、C312) : type 1B-3

C314 第12回国際鳥類保護会議記念
12th International Ornithological Congress

発行日：1960（昭和35）.5.24
Date of Issue：1960 (Showa 35).5.24

C314 トキ Japanese Crested Ibis

●特印 Special Date Stamp

東京 Tokyo

	**	●	FDC
C314 10 yen 赤・ピンク・灰 red, pink and gray	100	80	350

切手データ Basic Information

版式：グラビア輪転版	printing plate : rotary photogravure plate
目打：櫛型13½	perforation : comb perf. 13½
印面寸法：22.5×27mm	printing area size : 22.5×27mm
発行枚数：800万枚	number issued : 8 million
シート構成：20枚（横4×縦5）	pane format : 20 subjects (4x5)
製版・目打組合せ：タイプ1B-3	plate and perforation configuration : type 1B-3

C314 ● 定常変種 Constant Flaws

 A-pos.9
 A-pos.19
 B-pos.5
 B-pos.8
 B-pos.13
 B-pos.17
 C-pos.9
 D-pos.17

位置 position	特徴 feature	**	●
A-pos.9	脚の右に白点 white spot to right of leg	300	200
A-pos.19	"刷"と"局"の間に灰色点（銘版）gray dot between "刷" and "局" in the imprint		
B-pos.5	"1"の上方に白点 white dot above "1"		
B-pos.8	脚の上部に赤点 red dot on upper part of leg		
B-pos.13	翼に赤点 red dot on wing		
B-pos.17	上部マージンに赤点 red dot in top margin		
C-pos.9	"本"の下方に白点 white dot below "本"		
D-pos.17	首に赤点 red dot on neck		

C315 国際放送25年記念
25th Anniversary of Radio Japan

発行日：1960（昭和35）.6.1
Date of Issue：1960 (Showa 35).6.1

C315 地球をかこむ電波 Radio Waves Encircling Globe

●特印 Special Date Stamp

東京 Tokyo

	**	●	FDC
C315 10 yen 紅赤 carmine rose	60	50	260

切手データ Basic Information

版式：凹版速刷版	printing plate : high-speed intaglio printing
目打：櫛型13½	perforation : comb perf. 13½
印面寸法：27×22.5mm	printing area size : 27×22.5mm
発行枚数：800万枚	number issued : 8 million
シート構成：20枚（横5×縦4）	pane format : 20 subjects (5x4)
製版・目打組合せ：タイプ1C-2	plate and perforation configuration : type 1C-2

記念・特殊切手（額面「円位」時期、ローマ字なし）

C316 ハワイ官約移住75年記念　発行日：1960（昭和35）.8.20
75th Anniversary of Japanese Emigration to Hawaii　Date of Issue：1960 (Showa 35) .8.20

C316 サクラとパイナップルに虹
Cherry Blossoms, Pineapples and Rainbow

●特印 Special Date Stamp

横浜 Yokohama

	**	◉	FDC
C316 10 yen 多色 multicolored ·············	100	60	300

切手データ Basic Information	
版式：グラビア輪転版	printing plate : rotary photogravure plate
目打：櫛型13½	perforation : comb perf. 13½
印面寸法：27×22.5mm	printing area size : 27×22.5mm
発行枚数：800万枚	number issued : 8 million
シート構成：20枚（横5×縦4）	pane format : 20 subjects (5x4)
製版・目打組合せ：タイプ1C-2	plate and perforation configuration : type 1C-2

C316 ●定常変種 Constant Flaws

A-pos.2　B-pos.11　C-pos.3　D-pos.9

D-pos.16

位置 position	特徴 feature	**	◉
A-pos.2	桜の右に青点 blue dot to right of cherry blossoms	300	150
B-pos.11	"0"の右にピンク点 pink dot to right of "0"		
C-pos.3	年号"60"に緑点 green dot on "60"		
D-pos.9	虹の下方に青点 blue dot below rainbow		
D-pos.16	黄色の虹にピンク点 pink dot on yellow zone of rainbow		

C317 航空50年記念　発行日：1960（昭和35）.9.20
50th Anniversary of Japanese Aviation　Date of Issue：1960 (Showa 35) .9.20

C317 最初のファルマン機と1960年当時のジェット機
Henri Farman's Biplane and Jet Plane in 1960

●特印 Special Date Stamp

東京 Tokyo

	**	◉	FDC
C317 10 yen 青味灰・赤茶 chalky blue and brown ·············	60	50	270

切手データ Basic Information	
版式：グラビア輪転版	printing plate : rotary photogravure plate
目打：櫛型13½	perforation : comb perf. 13½
印面寸法：27×22,5mm	printing area size : 27×22.5mm
発行枚数：800万枚	number issued : 8 million
シート構成：20枚（横5×縦4）	pane format : 20 subjects (5x4)
製版・目打組合せ：タイプ1B-3	plate and perforation configuration : type 1B-3

位置 position	特徴 feature	**	◉
A-pos.9	ジェット機の下に白点 white dot under jet plane	200	100
A-pos.10	複葉機の下に白点 white dot under biplane		
B-pos.14	"本"の上に白点 white dot above "本"		
B-pos.16	"0"の下に白抜け、飛行機の下に白点、耳紙に茶点群 white spot under "0", white dot under biplane, brown dots in selvage		
C-pos.5	"郵便"の下方に白点 white dot below "郵便"		
C-pos.15	エンジンの下と"本"の下に白点 white dots under engine and under "本"		
D-pos.1	ジェット機の右に白点 white dot to right of jet plane		
D-pos.13	円状の白抜けと濃い点 circular white spot and dark dot		
D-pos.19	印面欠け nick in printing area		
D-pos.20	ジェット機の上方に白点 white dot above jet plane		

C317 ●定常変種 Constant Flaws

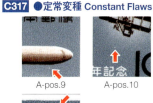

A-pos.9　A-pos.10

B-pos.14

B-pos.16

C-pos.5　C-pos.15　D-pos.1

D-pos.13　D-pos.19　D-pos.20

49

記念・特殊切手（額面「円位」時期、ローマ字なし）

C318-319　第49回列国議会同盟会議記念　発行日：1960（昭和35）.9.27
49th Congress of Inter-Parliamentary Union　Date of Issue : 1960 (Showa 35) .9.27

C318 議会の議席と英文による記念文字
Seat Plan of Diet Chamber and Commemorative Characters

C319 北斎の「赤富士」と国会議事堂
"Fine Breezy Day (Red Fuji)" by Katsushika Hokusai and Diet Building

● 特印
Special Date Stamp

東京 Tokyo

	**	●	FDC
C318 10 yen だいだい・暗い青 orange and indigo		50	40
C319 10 yen 赤茶・暗い青 red brown and blue		100	60
C318-319（2種連刷 pair）	150	100	400

切手データ Basic Information	
版式：グラビア輪転版	printing plate : rotary photogravure plate
目打：櫛型13½	perforation : comb perf. 13½
印面寸法：27×22.5mm	printing area size : 27×22.5mm
発行枚数：各800万枚	number issued : 8 million each
シート構成：各20枚（横5×縦4）	pane format : 20 subjects (5x4) each
製版・目打組合せ：タイプ1C-2	plate and perforation configuration : type 1C-2

C318 ● 定常変種 Constant Flaws

B-pos.2

A-pos.4

B-pos.6

C-pos.12

C-pos.13

D-pos.4

D-pos.7

D-pos.10

位置 position	特徴 feature	**	●
A-pos.4	議席の下に白点 white dot below seating area	200	100
B-pos.2	"5"の右方に白点、議席に青点と白点 white dot to right of "5", indigo dot and white dot on seats		
B-pos.6	議席に青の破線 blue dashed line on seats		
C-pos.12	議席に白点 white dot on seats		
C-pos.13	"4"の横棒が上にハネ error in "4"		
D-pos.4	下マージンに小さな青点 small blue dot in bottom margin		
D-pos.7	"便"の左に白点 white dot to left of "便"		
D-pos.10	"5"の右に白点 white dot to right of "5"		

C319 ● 定常変種 Constant Flaws

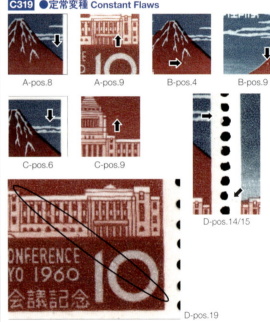

A-pos.8　A-pos.9　B-pos.4　B-pos.9

C-pos.6　C-pos.9

D-pos.14/15

D-pos.19

位置 position	特徴 feature	**	●
A-pos.8	稜線の右に白点 white dot to right of mountain ridge	300	150
A-pos.9	議事堂に茶点 brown dot on Diet Building		
B-pos.4	山の中腹に大きな白点 large white dot on mountainside		
B-pos.9	雲の下に白点、山の中腹に大きな白点 white dots under cloud, large white dot on mountainside		
C-pos.6	雲の上に白点 white dot above cloud		
C-pos.9	議事堂の右に白点 white dot to right of Diet Building		
D-pos.14/15	マージンに赤点2ヵ所 two red dots in margin		
D-pos.19	茶色の長い斜線 long brown diagonal line		

記念・特殊切手（額面「円位」時期、ローマ字なし）

C320 国際文通週間 / International Letter Writing Week 1960
発行日：1960（昭和35）.10.9　　Date of Issue：1960 (Showa 35).10.9

C320 歌川広重画「東海道五十三次 蒲原」 "The Fifty-three Stations of the Tokaido: Kanbara" by Utagawa Hiroshige

●特印 Special Date Stamp

蒲原 Kanbara

	**	●	FDC
C320 30 yen 多色 multicolored	3,500	1,000	5,000

切手データ Basic Information	
版式：グラビア輪転版	printing plate : rotary photogravure plate
目打：櫛型13	perforation : comb perf. 13
印面寸法：35.5×25mm	printing area size : 33.5×25mm
発行枚数：500万枚	number issued : 5 million
シート構成：20枚（横5×縦4）	pane format : 20 subjects (5x4)
製版・目打組合せ：タイプ1D-2	plate and perforation configuration : type 1D-2

C320 ● 定常変種 Constant Flaws

A-pos.1　A-pos.18　B-pos.5

B-pos.18　C-pos.3

C-pos.6

位置 position	特徴 feature	**	●
A-pos.1	"便"の上方に茶点 brown dot above "便"	8,000	3,000
A-pos.18	"本"の右上方に黒点、"便"の上方に白抜け black dot to upper right of "本", white spot above "便"		
B-pos.5	旅人の左に黒点 black dot to left of traveler		
B-pos.18	山頂の上に大きい黒点 large black dot above mountaintop		
C-pos.3	"NA"の下に黒点 black dot below "NA"		
C-pos.6	旅人の右に大きい黒点 large black dot to right of traveler		

C321 岡山天体物理観測所開所記念 / Opening of Okayama Astrophysical Observatory
発行日：1960（昭和35）.10.19　　Date of Issue：1960 (Showa 35).10.19

C321 観測所と瀬戸内海 Okayama Astrophysical Observatory and Seto Inland Sea

●特印 Special Date Stamp

鴨方 Kamogata

	**	●	FDC
C321 10 yen こい紫 bright violet	60	50	300

切手データ Basic Information	
版式：凹版速刷版	printing plate : high-speed intaglio printing
目打：櫛型13½	perforation : comb perf. 13½
印面寸法：33×22.5mm	printing area size : 33×22.5mm
発行枚数：800万枚	number issued : 8 million
シート構成：20枚（横5×縦4）	pane format : 20 subjects (5x4)
製版・目打組合せ：タイプ1C-2	plate and perforation configuration : type 1C-2

C322-323 第15回国体記念 / 15th National Athletic Meet
発行日：1960（昭和35）.10.23　　Date of Issue：1960 (Showa 35).10.23

C322 剣道 Kendo　C323 跳馬 Vault

●特印 Special Date Stamp

熊本 Kumamoto

	**	●	FDC
C322 10 yen 暗い青 dull blue	70	40	
C323 10 yen 暗い赤 rose violet	70	40	
C322-323 (2種連刷 pair)	150	160	550

切手データ Basic Information	
版式：ザンメルドック凹版	printing plate : sammeldruck intaglio printing
目打：櫛型13½	perforation : comb perf. 13½
印面寸法：22.5×27mm	printing area size : 22.5×27mm
発行枚数：各800万枚	number issued : 8 million each
シート構成：20枚（横4×縦5）	pane format : 20 subjects (4x5)
製版・目打組合せ：タイプ1B-3	plate and perforation configuration : type 1B-3

記念・特殊切手（額面「円位」時期、ローマ字なし）

C324 白瀬中尉南極探検50年記念　発行日：1960（昭和35）.11.29
50th Anniversary of Lieutenant Shirase's Antarctic Exploration　Date of Issue : 1960 (Showa 35) .11.29

C324　白瀬矗中尉と南極地図
Shirase Nobu and Map of Antarctica

●特印 Special Date Stamp

東京 Tokyo

	**	●	FDC
C324 10 yen くすみ赤・黒 fawn and black …… 70	50		400

切手データ Basic Information

版式：グラビア輪転版	printing plate : rotary photogravure plate
目打：櫛型13½	perforation : comb perf. 13½
印面寸法：27×22.5mm	printing area size : 27×22.5mm
発行枚数：800万枚	number issued : 8 million
シート構成：20枚（横5×縦4）	pane format : 20 subjects (5x4)
製版・目打組合せ：タイプ1C-2	plate and perforation configuration : type 1C-2

B-pos.15　B-pos.17

C-pos.7　C-pos.19

位置 position	特徴 feature	**	●
A-pos.2	鼻に黒点 black dot on nose	200	100
A-pos.4	"60"の下方に白点 white dot below "60"		
A-pos.18	顔に黒点 black dot on face		
B-pos.13	フードに黒点 black dot on hood		
B-pos.15	"19"の下方に白点 white dot below "19"		
B-pos.17	フードに黒の破線 black dashed line on hood		
C-pos.7	"日"と"本"の間に黒点、"本"の右下方に白点 black dot between "日" and "本", white dot to lower right of "本"		
C-pos.19	"1"の左に白点、(銘版)"造"の右下に小黒点 white dot to left of "1", small black dot to lower right of "造" in imprint		

C324 ●定常変種 Constant Flaws

A-pos.2　A-pos.4　A-pos.18　B-pos.13

C325-326 議会開設70年記念　発行日：1960（昭和35）.12.24
70th Anniversary of National Diet Building　Date of Issue : 1960 (Showa 35) .12.24

C325 国会議事堂と70年の年月を示す星
Diet Building and Stars Representing 70 Years

C326 第1回帝国議会の開院式
Opening of First Imperial Diet

●特印 Special Date Stamp

東京 Tokyo

	**	●	FDC
C325 10 yen 灰・暗い青 gray and dark blue	50	40	
C326 10 yen 暗い赤 carmine …………… 80		50	
C325-326 (2種 set of 2 values) …… 130		90	400

切手データ Basic Information

版式：C325=グラビア輪転版、C326=凹版速刷版	printing plate : C325=rotary photogravure plate, C326= high-speed intaglio printing
目打：櫛型13½	perforation : comb perf. 13½
印面寸法：27×22.5mm	printing area size : 27×22.5mm
発行枚数：各800万枚	number issued : 8 million each
シート構成：各20枚（横5×縦4）	pane format : 20 subjects (5x4) each
製版・目打組合せ：タイプ1C-2	plate and perforation configuration : type 1C-2

C325 ●定常変種 Constant Flaws

A-pos.20　B-pos.7　B-pos.14　C-pos.5

D-pos.3　D-pos.5　D-pos.19

位置 position	特徴 feature	**	●
A-pos.20	星の左下に白点 white dot at lower left of star	200	100
B-pos.7	"念"の下に白点 nick below "念"		
B-pos.14	第3コーナーマージンに灰色点 gray dot in SE corner margin		
C-pos.5	星の右下に白点 white dot to lower right of star		
D-pos.3	"本"の右上に白ぼやけ white blur to upper right of "本"		
D-pos.5	星の左に白点 white dot to left of star		
D-pos.19	議事堂の左に白点 white dot to left of Diet building		

1) 記念式典の延期により切手発行日が変更されたため、発行日が印刷された耳紙上部を断裁して発売された。

1) The issue date of this stamp was changed due to postponement of the commemorative ceremony. The stamps were therefore sold with the top selvage, on which the original issue date was printed, cut off.

記念・特殊切手（額面「円位」時期、ローマ字なし）

C327-338 花シリーズ Flower Series
発行年：1961（昭和36） Year of Issue：1961 (Showa 36)

 C327 スイセン Narcissus
 C328 ウメ Japanese Apricot
 C329 ツバキ Camellia
 C330 ヤマザクラ Mountain Cherry
 C331 ボタン Tree peony
 C332 ハナショウブ Japanese Iris

 C333 ヤマユリ Gold-Banded Lilly
 C334 アサガオ Morning Glory
 C335 キキョウ Balloon Flower
 C336 リンドウ Gentian
 C337 キク Chrysanthemum
 C338 サザンカ Sasanqua

●風景印 Pictorial Postmarks

 C327 四箇浦 Shikaura
 C328 水戸 Mito
 C329 大島 Oshima
 C330 吉野 Yoshino
 C331 大根島 Daikonshima
 C332 葛飾 Katsushika
C333 横須賀 Yokosuka
C334 入谷 Iriya
 C335 沼田 Numata
C336 坊中 Bochu
 C337 笠間 Kasama
C338 仁比山 Niiyama

	**	●	FDC
C327 10 yen (1961.1.30) 多色 multicolored	800	180	1,800
C328 10 yen (1961.2.28) 多色 multicolored	350	180	1,300
C329 10 yen (1961.3.20) 多色 multicolored	250	180	1,000
C330 10 yen (1961.4.28) 多色 multicolored	250	180	900
C331 10 yen (1961.5.25) 多色 multicolored	180	170	800
C332 10 yen (1961.6.15) 多色 multicolored	100	90	800
C333 10 yen (1961.7.15) 多色 multicolored	80	70	1,300
C334 10 yen (1961.8.1) 多色 multicolored	80	70	600
C335 10 yen (1961.9.1) 多色 multicolored	80	70	600
C336 10 yen (1961.10.2) 多色 multicolored	80	70	600
C337 10 yen (1961.11.1) 多色 multicolored	80	70	600
C338 10 yen (1961.12.1) 多色 multicolored	80	70	600
C327-338 (12種 set of 12 values)	2,410	1,400	

切手データ Basic Information

版式：グラビア輪転版	printing plate : rotary photogravure plate
目打：櫛型13½	perforation : comb perf. 13½
印面寸法：22.5×33mm	printing area size : 22.5×33mm
発行枚数：C327-C332：各800万枚、C333-C338：各1,000万枚	number issued : C327-C332: 8 million, C333-C338: 10 million
シート構成：20枚（横4×縦5）	pane format : 20 subjects (4x5)
製版・目打組合せ：C327-C328, C337-C338=タイプ1C-4, C329=タイプ1C-3, C330-C336=タイプ2B-1	plate and perforation configuration : C327-C328, C337-C338=type 1C-4, C329=type 1C-3, C330-C336=type 2B-1

●変則目打 Irregular Perforations

C330 C336

目打穴数 number of perforations into bottom selvage
2-3-2-3-2

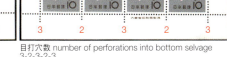
目打穴数 number of perforations into bottom selvage
3-2-3-2-3

C331

目打穴数 number of perforations into bottom selvage
1-2-1-2-1

目打穴数 number of perforations into bottom selvage
2-1-2-1-2

1) C330-C336の目打穿孔は連続櫛型目打のタイプ2B-1であるが、C330、C336には縦目打下側の針が1本多く植えられ、C331には1本少なく植えられたため変則目打になったものがある。

1) The perforation type of C330-C336 is the continuous comb perforation, type 2B-1. Irregular perforations, in which the numbers of perforations into the bottom selvage are irregular, are known for C330, C331 and C336.

記念・特殊切手（額面「円位」時期、ローマ字なし）

 C327
●定常変種 Constant Flaws

 A-pos.12　 B-pos.8

 B-pos.17　 C-pos.15

 D-pos.4　　D-pos.20

位置 position	特徴 feature	**	●
A-pos.12	濃い点 dark dot	3,000	500
B-pos.8	"す"の下に濃い点 dark dot below "す"		
B-pos.17	花びらの右に白点 white dot to right of petal		
C-pos.15	花びらの右上方に濃い点 dark dot to upper right of petal		
D-pos.4	花びらの上に濃い点 dark dot above petal		
D-pos.20	花びらに黒線、葉の右に濃い紫点 black line on petal, dark purple dot to right of leaf		

 C328
●定常変種 Constant Flaws

 A-pos.9　 A-pos.16　 B-pos.12　 B-pos.16

 B-pos.8　　 C-pos.11　C-pos.13

 D-pos.4　D-pos.18

位置 position	特徴 feature	**	●
A-pos.9	花びらに小さな黒点 small black dot on petal	1,000	500
A-pos.16	つぼみの左方に濃い茶点 dark brown dot to left of bud		
B-pos.8	花びらの左方に白点3個 three white dots to left of petal		
B-pos.12	"め"の右に白ぼやけ white blur to right of "め"		
B-pos.16	枝に大きな黒点 large black dot on branch		
C-pos.11	茎の間に白点2個 two white dots between stems		
C-pos.13	花びらの左方に白点 white dot to left of petal		
D-pos.4	枝の上方に破線状の白点 white dots above branch		
D-pos.18	中央の花びらに小さな黒点 small black dot on central petal		

 C329
●定常変種 Constant Flaws

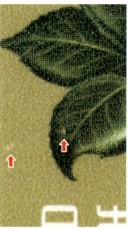

 B-pos.1　 B-pos.7　 B-pos.13

 C-pos.8　C-pos.17

A-pos.19　 D-pos.7　 D-pos.14

位置 position	特徴 feature	**	●
A-pos.19	白点と葉に緑抜け white dots to left of leaf, dot on leaf	700	500
B-pos.1	"ば"の右下に白点 white dot to lower right of "ば"		
B-pos.7	"日"の右下方に白点 white dot to lower right of "日"		
B-pos.13	"1"と"0"の間に白点 white dot between "1" and "0"		
C-pos.8	"つ"の下方に白点 white dot below "つ"		
C-pos.17	つぼみの下に白点 white dot below bud		
D-pos.7	葉に緑抜け dot on leaf		
D-pos.14	"1"と"0"の間に白点2個 two white dots between "1" and "0"		

記念・特殊切手（額面「円位」時期、ローマ字なし）

 C330
●定常変種 Constant Flaws

A-pos.2   A-pos.7   B-pos.17   B-pos.19

C-pos.8   D-pos.20

位置 position	特徴 feature	**	◉
A-pos.2	花びらの左に白点 white spot to left of petal	700	500
A-pos.7	葉の間に白点 white dot between leaves		
B-pos.17	葉の上方に白点 white dot above leaf		
B-pos.19	葉の上方に白点 white dot above leaf		
C-pos.8	"やま"の上方に白点 white dot above "やま"		
D-pos.20	"本"の上に白点 white dot above "本"		

 C332
●定常変種 Constant Flaws

A-pos.16   A-pos.17   A-pos.20   B-pos.17

C-pos.8   C-pos.11   D-pos.4

位置 position	特徴 feature	**	◉
A-pos.16	茎に緑抜け dot on stem	300	200
A-pos.17	"0"の右上方に白点 white dot to upper right of "0"		
A-pos.20	"0"の中に白点 white dot inside "0"		
B-pos.17	葉に緑抜け stain on leaf		
C-pos.8	花びらに緑点 green dot on petal		
C-pos.11	つぼみに緑点 green dot on bud		
D-pos.4	花びらに紫点 purple dot on petal		

 C331
●定常変種 Constant Flaws

A-pos.3   A-pos.17   A-pos.19   B-pos.6

B-pos.8   B-pos.18   C-pos.20   D-pos.6

位置 position	特徴 feature	**	◉
A-pos.3	葉に突起物 protrusion on leaf	600	500
A-pos.17	"ん"の右方に白点 white dot to right of "ん"		
A-pos.19	花びらに緑の横線 green horizontal line on petal		
B-pos.6	"ぽ"の上方に白線 white line above "ぽ"		
B-pos.8	"ん"の右方に白点 white dot to right of "ん"		
B-pos.18	花びらの右に白点 white dot to right of petal		
C-pos.20	"ぽ"の上方にあざ状の濃い点 dark bruise-like dot above "ぽ"		
D-pos.6	花びらに桃色点 pink dot on petal		

 C333
●定常変種 Constant Flaws

A-pos.2   A-pos.15   B-pos.7   B-pos.19

C-pos.11

C-pos.2

D-pos.14

D-pos.20

記念・特殊切手（額面「円位」時期、ローマ字なし）

位置 position	特徴 feature	**	●
A-pos.2	つぼみの左方に白点 white dot to left of bud	300	200
A-pos.15	右下角に白点 white dot in bottom right corner		
B-pos.7	葉の間に白点 white dot between leaves		
B-pos.19	葉の左下方に白点 white dot to lower left of leaf		
C-pos.2	花びらに小さい赤点 small red dot on petal		
C-pos.11	白抜け2ヵ所 two white spots		
D-pos.14	白点と白ぼやけ white dot and white blur		
D-pos.20	白点2ヵ所 two white dots		

 C334
●定常変種 Constant Flaws

A-pos.11

B-pos.16

B-pos.19

C-pos.10

C-pos.11

C-pos.17

位置 position	特徴 feature	**	●
A-pos.11	"郵"の右に白点 white dot to right of "郵"	300	200
B-pos.16	白点3ヵ所 three white dots		
B-pos.19	"便"の上方に白点 white dot above "便"		
C-pos.10	白点4ヵ所 four white dots		
C-pos.11	"10"の下方に白点 white dot below "10"		
C-pos.17	葉に緑抜け dot on leaf		

 C335
●定常変種 Constant Flaws

A-pos.12

A-pos.19

B-pos.12

B-pos.14

B-pos.18

C-pos.16

C-pos.18

D-pos.1

D-pos.12

D-pos.20

位置 position	特徴 feature	**	●
A-pos.12	"う"の下と花の間に白点 white dots below "う" and between flowers	300	200
A-pos.19	花びらの右方に白抜き white spot to right of petal		
B-pos.12	"よ"の下に白ぼかし white blur below "よ"		
B-pos.14	つぼみの右に白抜け white spot to right of bud		
B-pos.18	花びらの左に白点 white dot to left of petal		
C-pos.16	小さい紫点2ヵ所 two small purple dots		
C-pos.18	リング状の白抜け white ring		
D-pos.1	"日"の左下に白点 white dot to lower left of "日"		
D-pos.12	"き"の上方に白点 white dot above "き"		
D-pos.20	"き"の左方に白点 white dot to left of "き"		

 C336
●定常変種 Constant Flaws

A-pos.2

A-pos.6

B-pos.6

記念・特殊切手（額面「円位」時期、ローマ字なし）

B-pos.11

C-pos.1

C-pos.18

D-pos.1

D-pos.3

D-pos.16　D-pos.8　D-pos.19　D-pos.14

 C337
●定常変種 Constant Flaws

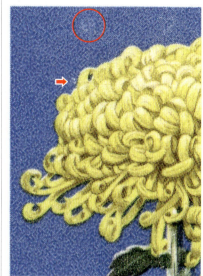
A-pos.2

A-pos.9

B-pos.6

C-pos.13

C-pos.15

D-pos.18

位置 position	特徴 feature	**	◉
A-pos.2	花の左上方に白ぼやけ、花の左に白点 white blur to upper left of flower, white dot to left of flower	300	200
A-pos.9	葉の右に小さな白点 small white dot to right of leaf		
B-pos.6	花びらの上に白点 white dot above petal		
C-pos.13	花びらの上方に白点 white dot above petal		
C-pos.15	"1"の上に白点 white dot above "1"		
D-pos.18	花びらの左に白点 white dot to left of petal		

位置 position	特徴 feature	**	◉
A-pos.2	葉に緑抜け、葉の下方に白点 dot on leaf, white dot below leaf	300	200
A-pos.6	"ど"の下に白点 white dot below "ど"		
B-pos.6	"り"の左上方に白曲線 white curve to upper left of "り"		
B-pos.11	葉の上方に白抜け white spot to upper right of leaf		
C-pos.1	"1"に小さな欠け small nick on "1"		
C-pos.18	"う"の右上方に大きな白点 large white dot to upper right of "う"		
D-pos.1	"日"の左下に白点 white dot to lower left of "日"		
D-pos.3	"ん"の上方に白点 white dot above "ん"		
D-pos.8	つぼみの左に白曲線 white curve to left of bud		
D-pos.14	葉の上方と下方に白点 white dots above leaf and below leaf		
D-pos.16	葉の上方に白点 white dot above leaf		
D-pos.19	葉の右上方に白点 white dot to upper right of leaf		

 C338
●定常変種 Constant Flaws

A-pos.4

B-pos.14　B-pos.17

57

記念・特殊切手（額面「円位」時期、ローマ字なし）

C-pos.10

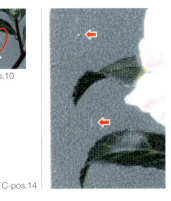
C-pos.14

C-pos.19

D-pos.8

位置 position	特徴 feature	**	●
A-pos.4	葉の左に白点、葉の下に大きなリタッチ white dot to left of leaf, large retouch below leaf	300	200
B-pos.14	葉の下に白点、"本"の上方に白ぼやけ white dot below leaf, white blur above "本"		
B-pos.17	葉の下に緑点、花びらに緑点 green dot below leaf, green dot on petal		
C-pos.10	葉の下に白点と大きな白ぼやけ white dot and large white blur below leaf		
C-pos.14	葉の上に白点、葉の下に白ぼやけ white dot above leaf, white blur below leaf		
C-pos.19	"さ"の上に白点 white dots above "さ"		
D-pos.8	葉の間に白ぼやけ2ヵ所 two white blurs between leaves		

C339 郵便90年記念
90th Anniversary of Postal Service
発行日：1961（昭和36）.4.20　　Date of Issue：1961 (Showa 36) .4.20

C339　前島密 Maejima Hisoka

●特印 Special Date Stamp

東京 Tokyo

C339 10 yen オリーブ・黒 olive and black　** 300　● 100　FDC 400

切手データ Basic Information	
版式：グラビア輪転版	printing plate：rotary photogravure plate
目打：櫛型13×13½	perforation：comb perf. 13×13½
印面寸法：18.5×22.5mm	printing area size：18.5×22.5mm
発行枚数：800万枚	number issued：8 million
シート構成：100枚（横10×縦10）	pane format：100 subjects (10x10)
製版・目打組合せ：タイプ2A	plate and perforation configuration：type 2A

C340 切手趣味週間
Philatelic Week 1961
発行日：1961（昭和34）.4.20　　Date of Issue：1961 (Showa 36) .4.20

C340「舞妓図屏風」"Screen of Dancers"

●特印 Special Date Stamp

東京 Tokyo

C340 ●定常変種 Constant Flaws

A-pos.1

B-pos.2

D-pos.1

D-pos.10

C-pos.9

C340 10 yen 多色 multicolored　** 150　● 150　FDC 900

位置 position	特徴 feature	**	●
A-pos.1	着物の裾に大きな青点 large blue dot on hem of kimono	400	400
B-pos.2	小さな黒点 small black dot		
C-pos.9	髪の近くに黒点2ヵ所 two black dots near hair		
D-pos.1	着物の右に小さな黒点 small black dot to right of kimono		
D-pos.10	"9"の左方に黒点 black dot to left of "9"		

切手データ Basic Information	
版式：グラビア輪転版	printing plate：rotary photogravure plate
目打：櫛型13½	perforation：comb perf. 13½
印面寸法：33×48mm	printing area size：33×48mm
発行枚数：1,000万枚	number issued：10 million
シート構成：10枚（横5×縦2）	pane format：10 subjects (5x2)
製版・目打組合せ：タイプ1C-1	plate and perforation configuration：type 1C-1

記念・特殊切手（額面「円位」時期、ローマ字なし）

C341 第52回国際ロータリー大会記念
52nd Rotary International Convention

発行日：1961（昭和36）.5.29
Date of Issue：1961 (Showa 36) .5.29

C341 ロータリーマークと世界の人々
Rotary Emblem and People of Various Countries

● 特印 Special Date Stamp

東京 Tokyo

	**	●	FDC
C341 10 yen 灰・だいだい gray and orange	50	50	280

切手データ Basic Information	
版式：ザンメル凹版	printing plate：sammeldruck intaglio printing
目打：櫛型13½	perforation：comb perf. 13½
印面寸法：27×22.5mm	printing area size：27×22.5mm
発行枚数：830万枚	number issued：8.3 million
シート構成：20枚（横4×縦5）	pane format：20 subjects (4x5)
製版・目打組合せ：タイプ1C-2	plate and perforation configuration：type 1C-2

C342 愛知用水通水記念
Inauguration of Aichi Irrigation System

発行日：1961（昭和36）.7.7
Date of Issue：1961 (Showa 36) .7.7

C342 蛇口、碍子（がいし）、小麦、歯車
Faucet, Insulator, Wheat stalk and Cogwheel

● 特印 Special Date Stamp

名古屋 Nagoya

	**	●	FDC
C342 10 yen にぶ青・暗い紫 violet and aquamarine	50	50	250

切手データ Basic Information	
版式：グラビア輪転版	printing plate：rotary photogravure plate
目打：櫛型13½	perforation：comb perf. 13½
印面寸法：22.5×27mm	printing area size：22.5×27
発行枚数：800万枚	number issued：8 million
シート構成：20枚（横4×縦5）	pane format：20 subjects (4x5)
製版・目打組合せ：タイプ1C-3	plate and perforation configuration：type 1C-3

C342 ● 定常変種 Constant Flaws

A-pos.1

A-pos.20

B-pos.13

C-pos.3

C-pos.4

D-pos.4

位置 position	特徴 feature	**	●
A-pos.1	蛇口の下方に白ぼやけ white blur below faucet	200	100
A-pos.20	第4コーナーマージンに紫点 purple dot at SW margin		
B-pos.13	霧状のぼやけ foggy blur		
C-pos.3	蛇口の左下方に白点 white spot below left of faucet		
C-pos.4	"1"の下に白点 white dot below "1"		
D-pos.4	"1"の下に紫点 purple dot below "1"		

C343 日本標準時制定75年記念
75th Anniversary of Japan Standard Time

発行日：1961（昭和36）.7.12
Date of Issue：1961 (Showa 36) .7.12

C343 太陽と日本標準時子午線
Sun and Japan Standard Time Meridian

● 特印 Special Date Stamp

明石 Akashi

	**	●	FDC
C343 10 yen くすみ赤・黒 fawn and black	50	50	250

切手データ Basic Information	
版式：グラビア輪転版	printing plate：rotary photogravure plate
目打：櫛型13½	perforation：comb perf. 13½
印面寸法：27×22.5mm	printing area size：27×22.5mm
発行枚数：800万枚	number issued：8 million
シート構成：20枚（横5×縦4）	pane format：20 subjects (5x4)
製版・目打組合せ：タイプ1C-2	plate and perforation configuration：type 1C-2

記念・特殊切手（額面「円位」時期、ローマ字なし）

C343 ●定常変種 Constant Flaws

A-pos.1

A-pos.9

A-pos.15

B-pos.15

C-pos.17

C-pos.19

D-pos.14

D-pos.19

位置 position	特徴 feature	**	●
A-pos.1	"明石"の下方に大きな黒点 large black dot below "明石"	200	100
A-pos.9	地球の左下方に白点 white dot to lower left of globe		
A-pos.15	地球の右上に黒点 black dot on upper right part of globe		
B-pos.15	"年"に文字欠け defect in "年"		
C-pos.17	"日"の上に白点 white dot above "日"		
C-pos.19	"日"に黒点 black dot on "日"		
D-pos.14	"本"の下方と地球の右上方に各大きな白点 large white dots below "本" and to upper right of globe		
D-pos.19	"明石"の右上方に大きな黒点群 large black dots to upper right of "明石"		

C344-345 第16回国体記念
16th National Athletic Meet

発行日：1961（昭和36）.10.8
Date of Issue : 1961 (Showa 36) .10.8

C344 鉄棒
horizontal bar

C345 漕艇
Rowing

●特印 Special Date Stamp

秋田 Akita

		**	●	FDC
C344	5 yen 暗い青緑 blue green	60	40	
C345	5 yen 暗い青紫 ultramarine	60	40	
C344-345	(2種連刷 pair)	150	160	450

版式：ザンメル凹版	printing plate : sammeldruck intaglio printing
目打：櫛型13½	perforation : comb perf. 13½
印面寸法：27×22.5mm	printing area size : 27×22.5mm
発行枚数：各1,000万枚	number issued : 10 million each
シート構成：20枚（横5×縦4）	pane format : 20 subjects (5×4)
製版・目打組合せ：タイプ1C-4	plate and perforation configuration : type 1C-4

C346 国際文通週間
International Letter Writing Week 1961

発行日：1961（昭和36）.10.8
Date of Issue : 1961 (Showa 36) .10.8

C346 歌川広重画「東海道五十三次 箱根」
"The Fifty-three Stations of the Tokaido: Hakone" by Utagawa Hiroshige

●特印 Special Date Stamp

箱根町 Hakone-Machi

	**	●	FDC
C346 10 yen 多色 multicolored	1,500	1,200	3,500

版式：グラビア輪転版	printing plate : rotary photogravure plate
目打：櫛型13	perforation : comb perf. 13
印面寸法：35.5×25mm	printing area size : 35.5×25mm
発行枚数：500万枚	number issued : 5 million
シート構成：20枚（横5×縦4）	pane format : 20 subjects (5×4)
製版・目打組合せ：タイプ1D-2	plate and perforation configuration : type 1D-2

C346 ●定常変種 Constant Flaws

A-pos.2

C-pos.13

C-pos.17

位置 position	特徴 feature	**	●
A-pos.2	"9"の上に黒点 black dot above "9"	4,000	3,000
C-pos.13	山頂の右に大きい黒点 large black dot to right of mountaintop		
C-pos.17	題名の下方に黒点 black dot below title		

記念・特殊切手（額面「円位」時期、ローマ字なし）

```
第5次 5th issue 1963 (Showa 38) .11.11.    **     ●     FDC
C359  5 yen + 5 yen 暗い青 dark blue ……… 50    50
C360  5 yen + 5 yen オリーブ緑 olive ………… 50   50
C361  5 yen + 5 yen 暗い灰 black …………… 50    50
C362  5 yen + 5 yen 暗い赤紫 claret……… 50    50
      C359-362 (4種 set of 4 values)  ・200  200   600
第6次 6th issue 1964 (Showa 39) .6.23
C363  5 yen + 5 yen くすみ緑味青 greenish blue 50  50
C364  5 yen + 5 yen 暗い赤紫 rose claret……… 50   50
C365  5 yen + 5 yen 暗い黄茶 deep olive ……50    50
C366  5 yen + 5 yen くすみ青紫 bluish violet ・50  50
      C363-366 (4種 set of 4 values)  ・200  200   600
      C347-366 (20種 set of 20 values) ……1,810  2,500
```

切手データ Basic Information

版式：凹版速刷版	printing plate : high-speed intaglio printing
目打：櫛型 13½	perforation : comb perf. 13½
印面寸法：27×27mm	printing area size : 27×27mm
発行枚数：	number issued :
C347-349=各400万枚	C347-349=4 million each
C350-352=各450万枚	C350-3522=4.5 million each
C353-355=各500万枚	C353-355=5 million each
C356-358=各800万枚	C356-358=8 million each
C359-366=各1,400万枚	C359-366=14 million each
シート構成：各20枚（横4×縦5）	pane format : 20 subjects (4x5) each
製版・目打組合せ：	plate and perforation configuration :
C347-355、C359=タイプ1C-4	C347-355, C359=type 1C-4
C356-358=タイプ1C-3	C356-358=type 1C-3
C360=タイプ1C-4、タイプ1C-5	C360=type 1C-4 and type 1C-5
C361-366=タイプ1C-5	C361-366=type 1C-5

1) この時期の凹版印刷は小型輪転凹版印刷機（ザンメル凹版）によるが、東京1964オリンピック競技大会は小型シートを除き凹版速刷機による印刷である。
2) C360には上抜け目打の窓口シートと上下抜け目打の窓口シートがあり、後者には櫛歯上向きと櫛歯下向きがある。これは、製版・目打組合せタイプ1C-4（櫛歯上向き）、タイプ1C-5（櫛歯下向き）が併用されたことによる。

1) While the intaglio printed stamps in this period were usually printed on the small rotary intaglio (sammeldruck intaglio) press, these "Tokyo 1964 Olympic Games" stamps were printed on the high-speed intaglio press except for the souvenir sheets.
2) There are two types of perforation for C360: perforated through the upper selvage (1C-4), and perforated through the bottom selvage (1C-3). For the latter, there is a further variation with both upper comb perforations and bottom comb perforations. This situation occurred when the two plate and

●目打の抜け方 Position of Perforations Through Selvages

type 1C-3
下抜け・上下抜け
perforation through bottom selvage or top and bottom selvages

type 1C-4
上抜け・上下抜け
perforation through top selvage or top and bottom selvages

type 1C-5
上下抜け
perforation through top and bottom selvages

perforation configuration types, 1C-4 and 1C-5 were used in combination.

●目打の継ぎ目 Comb Perforator Joints

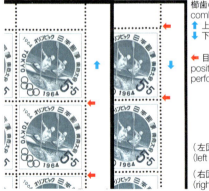

櫛歯の向き direction of comb perforator teeth
⬆ 上向き upward
⬇ 下向き downward

⬅ 目打針の継ぎ目 position of joints in perforator teeth

（左図）上向き
(left Fig) upward

（右図）下向き
(right Fig) downward

C367-372 東京1964オリンピック競技大会（寄附金付）小型シート
Tokyo 1964 Olympic Games (Semi-Postal), Souvenir Sheet

発行日：1964（昭和39）.8.20
Date of Issue : 1964 (Showa 39) .8.20

C367 第1次 1st issue

C368 第2次 2nd issue

●不整目打 irregular perforations

目打ずれ mis-aligned perforations

記念・特殊切手（額面「円位」時期、ローマ字なし）

C369 第3次　3rd issue

C370 第4次　4th issue

二重目打 double perforation

●目打もれ missing perforations

C371 第5次　5th issue

C372 第6次　6th issue

		**	●	FDC
C367	第1次　1st issue (C347-349) ………	1,100	1,200	1,800
	va. 目打もれ missing perforations …	70,000	-	
C368	第2次　2nd issue (C350-352) ………	700	800	1,800
	va. 目打もれ missing perforations …	70,000	-	
C369	第3次　3rd issue (C353-355) ………	600	700	1,800
	va. 目打もれ missing perforations …	70,000	-	
C370	第4次　4th issue (C356-358) ………	1,300	1,400	1,800
	va. 目打もれ missing perforations …	70,000	-	
C371	第5次　5th issue (C359-362) ………	1,300	1,400	1,800
	va. 目打もれ missing perforations …	70,000	-	
C372	第6次　6th issue (C363-366) ………	1,300	1,400	1,800
C367-372 (6種 set of 6 values) ………		6,300	6,900	

切手データ Basic Information	
版式：凹版	printing plate: intaglio printing
シート寸法：134.5×60mm	sheet size: 134.5×60mm
発行枚数： C367-369=各200万枚、 C370-372=各150万枚	number issued: C367-369=2 million each, C370-372=1.5 million each

C373　国立国会図書館新庁舎開館記念
Opening of New National Diet Library

発行日：1961（昭和36）.11.1　Date of Issue：1961 (Showa 36).11.1

C373　新庁舎とマーク
National Diet Library and Symbol

●特印
Special Date Stamp

東京 Tokyo

	**	●	FDC
C373 10 yen 紫味青・金 deep ultramarine and gold ………	50	50	250

切手データ Basic Information	
版式：グラビア輪転版	printing plate: rotary photogravure plate
目打：櫛型13½	perforation: comb perf. 13½
印面寸法：27×22.5mm	printing area size: 27×22.5mm
発行枚数：800万枚	number issued: 8 million
シート構成：20枚（横5×縦4）	pane format: 20 subjects (5×4)
製版・目打組合せ：タイプ1C-2	plate and perforation configuration: type 1C-2

C373 ●定常変種 Constant Flaws

A-pos.2

A-pos.7

B-pos.17

B-pos.19

C-pos.1　C-pos.16

D-pos.1

D-pos.6

位置 position	特徴 feature	**	●
A-pos.2	"1"の左上方に白点, "1"の下に白ぼやけ white dot to upper left of "1", white blur below "1"	200	100
A-pos.7	書籍の中央に青点 blue dot in center of book		
B-pos.17	図書館の右に白点 white dot to right of library		
B-pos.19	"9"の下方とマークの中と図書館の下に各白点 white dots below "9", inside symbol and below library		
C-pos.1	"6"の下方に大きな白点 large white dot below "6"		
C-pos.16	"9"の下方に大きな白点群 large white dots below "9"		
D-pos.1	書籍の上に白点 white dot above book		
D-pos.6	"舎"の左上方に白点 white dot to upper left of "舎"		

記念・特殊切手（額面「円位」時期、ローマ字なし）

C374-377 年中行事シリーズ　Seasonal Folklore Series
発行年：1962（昭和37）-1963（昭和38）　year of issue : 1962 (Showa37) -1963 (Showa38)

C374 ひなまつり　Hinamatsuri Festival
C375 たなばた　Tanabata Festival
C376 七五三　Shichi-go-san Festival
C377 節分　Setsubun Festival

●風景印 Pictorial Postmarks

C374 中京 Chukyo　C375 仙台 Sendai　C376 渋谷 Shibuya　C377 成田 Narita

	**	●	FDC
C374 10 yen 多色 multicolored (1962.3.3)	140	100	600
C375 10 yen 多色 multicolored (1962.7.7)	50	50	500
C376 10 yen 多色 multicolored (1962.11.15)	50	50	600
C377 10 yen 多色 multicolored (1963.2.3)	50	50	500
C374-377 (4種 Set of 4 Values)	290	250	

切手データ Basic Information

版式：グラビア輪転版	printing plate : rotary photogravure plate
目打：櫛型13½	perforation : comb perf. 13½
印面寸法：22.5×27mm	printing area size : 22.5×27mm
発行枚数：C374=800万枚、C375-377=各1,000万枚	number Issued : C374=8 million, C375-377=10 million each
シート構成：20枚（横4×縦5）	pane format : 20 subjects (4x5)
製版・目打組合せ：タイプ2B-1	plate and perforation configuration : type 2B-1

C374 ●定常変種 Constant Flaws

A-pos.8

位置 position	特徴 feature	**	●
A-pos.6	枝先に赤点 red dot on branch tip	400	200
A-pos.8	頭の上方に黒点 black dot above head		
A-pos.19	頬に黒点 black dot on cheek		
B-pos.6	"便"の右方に赤点 red dot to right of "便"		
B-pos.12	女雛の頭後方に黒点 black dot behind female Hina doll's head		
B-pos.17	"便"の下方に黒点 black dot below "便"		
C-pos.3	髪飾りの左方に白抜け white spot to left of hair ornament.		
C-pos.8	頭の右に黒色抜け spot to right of head		
C-pos.9	男雛の背中に小さい黒点 small black dot on back of male Hina doll		
D-pos.12	女雛の頭前方に黒点 black spot in front of female Hina doll's head		
D-pos.20	男雛に前方に黒点 black dot in front of male Hina doll		

A-pos.6　A-pos.19　B-pos.6　B-pos.12

B-pos.17　C-pos.3　C-pos.8　C-pos.9

D-pos.12　D-pos.20

C375 ●変則目打 irregular perforation

目打穴数 number of perforations into bottom selvage 2-1-2-1-2

目打穴数 number of perforations into bottom selvage 1-2-1-2-1

記念・特殊切手（額面「円位」時期、ローマ字なし）　記念・特殊

1) 年中行事シリーズの目打穿孔は連続櫛型目打のタイプ2B-1であるが、C375は縦目打下側の針が1本少なく植えられたため変則目打が生じたものがある。
1) The perforation of "Season's Folklore Series" is the continuous comb perforation, type 2B-1. An irregular perforation, in which one perforation pin is missing in the bottom selvage, is known for C375.

● 定常変種 Constant Flaws

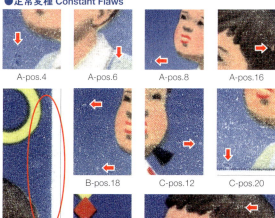

A-pos.15

B-pos.5

B-pos.16

C-pos.1

C-pos.20

D-pos.10

D-pos.19

位置 position	特徴 feature	**	●
A-pos.4	印面右側に赤点2ヵ所 two red dots on right margin	200	100
A-pos.15	"郵"の上方と"便"の下方に各紫点 each purple dots above "郵" and below "便"		
B-pos.5	印面右側に黒の突起物 black protrusion in right margin		
B-pos.16	紫点2ヵ所 two purple dots		
C-pos.1	唇の下に赤点 red dot below lip		
C-pos.20	"五三"の下方に赤点 red spot below "五三"		
D-pos.10	額に赤点 red dot on forehead		
D-pos.19	"日"の下に極小の赤点 tiny red dot below "日"		

位置 position	特徴 feature	**	●
A-pos.4	顎の下に濃い青点群 dark blue dots below chin	200	100
A-pos.6	襟元に青点 blue dot on collar		
A-pos.8	顎の左下方に青点 blue dot to under left of chin		
A-pos.16	頭に白抜け white spot on head		
B-pos.13	白点と白ぼやけ white dot and white blur		
B-pos.18	顔の左方に白点2ヵ所 two white dots to left of face		
C-pos.12	顔の右方に白点 white dot to right of face		
C-pos.20	リタッチの青点群 retouched blue dots		
D-pos.1	白キズと霧状の白ぼやけ white scratches and misty white blur		
D-pos.3	頭に濃い黒点、耳の右に青点 dark black dot on head, blue dot to right of ear		

C376
● 定常変種 Constant Flaws

A-pos.4

C377 ● 定常変種 Constant Flaws

A-pos.4

B-pos.5

C-pos.15

D-pos.15

位置 position	特徴 feature	**	●
A-pos.4	"日"の上とズボンに各白点 white dots above "日" and on pants	200	100
B-pos.5	"日"の下方に白点2ヵ所 two white dots below "日"		
C-pos.15	桝の左下方に白点 white dot to below left of square wooden cup		
D-pos.15	手の甲に赤点 red dot on hand		

65

記念・特殊切手（額面「円位」時期、ローマ字なし）

C378 切手趣味週間　Philatelic Week 1962
発行日：1962（昭和37）.4.20　date of issue : 1962 (Showa37) .4.20

●特印 Special Date Stamp

東京 Tokyo

C378 狩野長信画「花下遊楽図」
" Merry-Making under Flowers "
by Kano Naganobu

C378●定常変種 Constant Flaws

| A-pos.2 | A-pos.3 | A-pos.6 | B-pos.4 |

B-pos.9　　C-pos.5

C-pos.6　　　　C-pos.5

D-pos.3　　D-pos.9　　D-pos.10

	**	●	FDC
C378 10 yen 多色 multicolored	140	140	800

切手データ Basic Information	
版式：グラビア輪転版	printing plate : rotary photogravure plate
目打：櫛型13½	perforation : comb perf. 13½
印面寸法：33×48mm	printing area size : 33×48mm
発行枚数：1,000万枚	number issued : 10 million
シート構成：10枚（横5×縦2）	pane format : 10 subjects (5x2)
製版・目打組合せ：タイプ1C-2	plate and perforation configuration : type 1C-2

位置 position	特徴 feature	**	●
A-pos.2	扇子の右上方に大きな白抜け large white spot to upper right of fan	400	400
A-pos.3	足の前方に白点4個 four white dots in front of leg		
A-pos.6	頭の右上方に白点 white dot to upper right of head		
B-pos.4	"本"の下方に大きな白抜け、腕の上方に白抜け large white spot below "本", white spot above arm		
B-pos.9	"日"の下方に濃い茶 dark brown blur below "日"		
C-pos.5	"便"の下に白抜き、頭の右方に大きな白ぼやけ white spot below "便", large white blur to right of head.		
C-pos.6	"日"の下に白点と赤点、"0"の枠線欠け white dot and red dot below "日", nick on "0"		
D-pos.3	頭の右方に白点、右マージンに茶点 white dot to right of head, brown dot at right margin		
D-pos.9	扇子の右に大きな茶点 large brown dot to right of fan		
D-pos.10	左足の上に大きな黒点 large black dot on foot		

C379 北陸トンネル開通記念　Opening of Hokuriku Railway Tunnel
発行日：1962（昭和37）.6.10　Date of Issue：1962 (Showa 37) .6.10

C379 北陸トンネルと特急「白鳥」
Hokuriku Railway Tunnel and Limited Express "Hakucho"

●特印 Special Date Stamp

敦賀 Tsuruga

C379●定常変種 Constant Flaws

A-pos.11

A-pos.13

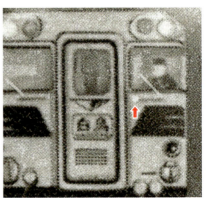

A-pos.18

	**	●	FDC
C379 10 yen 暗いオリーブ olive gray	150	80	550

切手データ Basic Information	
版式：グラビア輪転版	printing plate : rotary photogravure plate
目打：櫛型13½	perforation : comb perf. 13½
印面寸法：27×22.5mm	printing area size : 27×22.5mm
発行枚数：800万枚	number issued : 8 million
シート構成：20枚（横5×縦4）	pane format : 20 subjects (5x4)
製版・目打組合せ：タイプ1C-2	plate and perforation configuration : type 1C-2

記念・特殊切手（額面「円位」時期、ローマ字なし）

B-pos.5　　B-pos.13

B-pos.18　　C-pos.3

C-pos.18　　D-pos.2

D-pos.8　　D-pos.19　　　　C-pos.11

位置 position	特徴 feature	**	●
A-pos.11	車両の右に濃い点 dark dot to right of train	400	200
A-pos.13	"通"の上に濃い点 dark dot above "通"		
A-pos.18	車両に円状の白抜け circular white spot on train		
B-pos.5	光線に小さな点2個 two small dots on ray		
B-pos.13	"0"の右下に白傷 white scratch to lower right of "0"		
B-pos.18	車両に白点 white dot on train		
C-pos.3	"1"に小さな点 small dot on "1"		
C-pos.11	レールが切断、"本"と"郵"の間に小さい点 break in rail, small dot between "本" and "郵"		
C-pos.18	光線に小さな点 small dot on ray		
D-pos.2	"本"に黒点 black dot on "本"		
D-pos.8	"ト"の右に白点 white dot to right of "ト"		
D-pos.19	"郵"の下方に白点 white dot below "郵"		

C380　アジア・ジャンボリー記念　Asian Boy Scout Jamboree

発行日：1962（昭和37）.8.3
Date of Issue：1962 (Showa 37) .8.3

C380 ボーイスカウトの制帽とアジア地図
Boy Scout Hat and Asia Map

●特印 Special Date Stamp

御殿場 Gotenba

		**	●	FDC
C380 10 yen 多色 multicolored		50	40	350

切手データ Basic Information	
版式：グラビア輪転版	printing plate : rotary photogravure plate
目打：櫛型13½	perforation : comb perf. 13½
印面寸法：22.5×27mm	printing area size : 22.5×27mm
発行枚数：1,000万枚	number issued : 10 million
シート構成：20枚（横4×縦5）	pane format : 20 subjects (4×5)
製版・目打組合せ：タイプ2B-1	plate and perforation configuration : type 2B-1

1) 上下の耳紙に目打穴が2個飛び出す二連2型で穿孔されたが、二連の櫛歯端の縦打上側の目打針を1本少なく植えたため、変則目打になったものがある。

1) The perforation of this stamp is side-feed double-row comb type with the double-pin perforation into both upper and bottom selvages. Since in some cases one pin was removed at the upper side, the irregular perforation varieties with both single- and double-pin perforations into the upper selvage are shown.

C380　●定常変種 Constant Flaws

A-pos.5　　B-pos.5　　C-pos.14

C-pos.16　　D-pos.2　　D-pos.20

位置 position	特徴 feature	**	●
A-pos.5	"リ"の上方に白点 white dot above "リ"	200	100
B-pos.5	"R"の左右に小さな黒点 small black dot to left of "R"		
C-pos.14	"0"の上方に白点 white dot above "0"		
C-pos.16	帽子上部に黒色もれ top of hat faded		
D-pos.2	"ア"の上方に小さな黒点 small black dot above "ア"		
D-pos.20	"0"の右上方に小さな黒点 small black dot to upper right of "0"		

●変則目打 Irregular Perforation 上耳紙 upper selvage

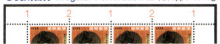

目打穴数 number of perforations into upper selvage 1-2-1-2-1

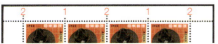

目打穴数 number of perforations into upper selvage 2-1-2-1-2

記念・特殊切手（額面「円位」時期、ローマ字なし）

C381 若戸大橋開通記念　発行日：1962（昭和37）.9.26
Opening of Wakato Bridge　Date of Issue : 1962 (Showa 37) .9.26

C381
若戸大橋全景
Wakato Bridge

●特印
Special Date Stamp

若松 Wakamatsu

	**	●	FDC
C381 10 yen 赤 rose red	100	40	400

切手データ Basic Information	
版式：凹版速刷版	printing plate : high-speed intaglio printing
目打：櫛型13½×13	perforation : comb perf. 13½×13
印面寸法：22.5×40mm	printing area size : 22.5×40mm
発行枚数：900万枚	number issued : 9 million
シート構成：20枚（横5×縦4）	pane format : 20 subjects (5x4)
製版・目打組合せ：タイプ1D-5	plate and perforation configuration : type 1D-5

C382 国際文通週間　発行日：1962（昭和37）.10.7
International Letter Week 1962　Date of Issue : 1962 (Showa 37) .10.7

C382 歌川広重画
「東海道五十三次
日本橋」
"The Fifty-
three Stations
of the Tokaido:
Nihonbashi"
by Utagawa
Hiroshige

●特印
Special Date Stamp

日本橋 Nihonbashi

	**	●	FDC
C382 40 yen 多色 multicolored	1,200	1,000	3,200

切手データ Basic Information	
版式：グラビア輪転版	printing plate : rotary photogravure plate
目打：櫛型13	perforation : comb perf. 13
印面寸法：35.5×25mm	printing area size : 33.5×25mm
発行枚数：500万枚	number issued : 5 million
シート構成：20枚（横5×縦4）	pane format : 20 subjects (5x4)
製版・目打組合せ：タイプ1D-1	plate and perforation configuration : type 1D-1

C382 ● 定常変種 Constant Flaws

A-pos.20

B-pos.15

C-pos.15

C-pos.18

D-pos.9

D-pos.10

位置 position	特徴 feature	**	●
A-pos.20	左側の縦帯に黒点 black dot on left vertical line	4,000	2,000
B-pos.15	題名の下に黒点 black dot below title		
C-pos.15	犬の足元に大きな白抜け large white spot beside dog's feet		
C-pos.18	"日"の上に青点 blue dot above "日"		
D-pos.9	題名の下方に2本の朱線 two vermilion lines below title		
D-pos.10	印章に斜線 diagonal line on seal		

C383-384 第17回国体記念　発行日：1962（昭和37）.10.21
17th National Athletic Meet　Date of Issue : 1962 (Showa 37) .10.21

C383 ライフル射撃
Rifle and Pistol

C384 ソフトボール
Softball

●特印
Special Date Stamp

岡山 Okayama

	**	●	FDC
C383 5 yen 暗い紫 brown violet	50	40	
C384 5 yen 暗い青 bluish black	50	40	
C383-384 (2種連刷 pair)	110	160	350

切手データ Basic Information	
版式：ザンメル凹版	printing plate : sammeldruck intaglio printing
目打：櫛型13½	perforation : comb perf. 13½
印面寸法：22.5×27mm	printing area size : 22.5×27mm
発行枚数：各1,000万枚	number issued : 10 million each
シート構成：20枚（横4×縦5）	pane format : 20 subjects (4x5)
製版・目打組合せ：タイプ1C-4	plate and perforation configuration : type 1C-4

記念・特殊切手（額面「円位」時期、ローマ字なし）

C385 北九州市発足記念　発行日：1963（昭和38）.2.10
Inauguration of Kitakyushu City　Date of Issue：1963 (Showa 38) .2.10

C385 地図、ハト、汽船、煙突
Map of City, Birds, Ship and Factory

●特印
Special Date Stamp

北九州 Kitakyushu

	**	◎	FDC
C385 10 yen 暗い茶 chocolate	50	30	250

切手データ Basic Information
版式：グラビア輪転版	printing plate：rotary photogravure plate
目打：櫛型13½	perforation：comb perf. 13½
印面寸法：27×22.5mm	printing area size：27×22.5mm
発行枚数：1,000万枚	number issued：10 million
シート構成：20枚（横5×縦4）	pane format：20 subjects (5×4)
製版・目打組合せ：タイプ2B-2	plate and perforation configuration：type 2B-2

C385 ●定常変種 Constant Flaws

A-pos.8　　A-pos.16　　A-pos.19

B-pos.4　　B-pos.19

C-pos.2　　C-pos.18　　C-pos.19

D-pos.6　　D-pos.16　　D-pos.19 正常 normal

位置 position	特徴 feature	**	◎
A-pos.8	マストの左に白抜け white spot to left of mast	200	100
A-pos.16	地図に濃い点 dark dot on map		
A-pos.19	"3"の右下に白点 white dot to lower right of "3" 銘版の上に線 line above imprint		
B-pos.4	地図に大きな濃い点2個 two large dark dots on map		
B-pos.19	汽船の煙突に太い線 thick line on steamer 銘版の上に線3ヵ所 three lines above imprint		
C-pos.2	"0"の上に白点、船尾に傷3ヵ所 white spot above "0", three scratches on stern		
C-pos.18	"3"に茶点 brown dot on "3"		
C-pos.19	銘版の上に線2ヵ所 two lines above imprint		
D-pos.6	地図に大きな濃い点 large dark dot on map		
D-pos.16	地図の左上に白点、"19"の左上方に白いもや white dot to upper left of map, white haze to upper left of "19"		

C386 飢餓救済運動　発行日：1963（昭和38）.3.21
Freedom from Hunger　Date of Issue：1963 (Showa 38) .3.21

C386 地球と飢餓救済運動のマーク
Globe and Emblem of Freedom from Hunger

●特印
Special Date Stamp

東京 Tokyo

	**	◎	FDC
C386 10 yen 暗い緑 dark green	50	30	250

切手データ Basic Information
版式：グラビア輪転版	printing plate：rotary photogravure plate
目打：櫛型13½	perforation：comb perf. 13½
印面寸法：33×22.5mm	printing area size：33×22.5mm
発行枚数：1,100万枚	number issued：11 million
シート構成：20枚（横5×縦4）	pane format：20 subjects (5×4)
製版・目打組合せ：タイプ2B-2	plate and perforation configuration：type 2B-2

C386 ●定常変種 Constant Flaws

A-pos.17　　A-pos.19　　B-pos.12

B-pos.18　　C-pos.7

C-pos.16　　D-pos.2

D-pos.11

記念・特殊切手（額面「円位」時期、ローマ字なし）

位置 position	特徴 feature	**	●
A-pos.17	南米大陸の下方に大きなリタッチ large retouch below South American continent	200	100
A-pos.19	"大"の左方に緑点2個、"造"の上に緑の斜線（銘版） two green dots to left of "大", green diagonal line above "造" of imprint		
B-pos.12	東南アジアの右に大きな白点 large white dot to right of Southeast Asia		
B-pos.18	豪州の右とインドの右に白点 white dots to right of Australia and India		
C-pos.7	"1"の中に小さな点 small dot on "1"		
C-pos.16	インドの右とアフリカ大陸の右に白点 white dots to right of India and African continent		
D-pos.2	"運"の上方に白点 white dot avobe 運		
D-pos.11	印面左側に多数の白点 many white dots at left of printing area		

C387 切手趣味週間 / Philatelic Week 1963

発行日：1963（昭和38）.4.20
Date of Issue：1963 (Showa 38) .4.20

●特印 Special Date Stamp
東京 Tokyo

C387 本多平八郎姿絵屏風（通称「千姫」）
Screen painting, Hondaheihachiro sugatae byobu (commonly known as "Senhime")

		**	●	FDC
C387 10 yen 多色 multicolored		60	60	550

切手データ Basic Information	
版式：グラビア輪転版	printing plate : rotary photogravure plate
目打：櫛型13½	perforation : comb perf. 13½
印面寸法：33×48mm	printing area size : 33×48mm
発行枚数：1,450万枚	number issued : 14.5 million
シート構成：10枚（横5×縦2）	pane format : 10 subjects (5x2)
製版・目打組合せ：タイプ1C-2	plate and perforation configuration : type 1C-2

C387 ●定常変種 Constant Flaws

A-pos.6   A-pos.9   B-pos.9

B-pos.9

C-pos.1

C-pos.9

D-pos.4

D-pos.5   D-pos.6   D-pos.9   D-pos.10

位置 position	特徴 feature	**	●
A-pos.6	"0"に線切れ line break on "0"	200	200
A-pos.9	"1"の左下方に茶点 brown dot to lower left of "1"		
	"大"の右下に黒点 black dot to lower right of "大" of imprint		
B-pos.9	足の下方にかすれと濃い点、足元の左に白点 faint and dark dot below foot, white dot to left of foot		
	"蔵"の先端にハネかすれ splash scratch at tip of "蔵" of imprint		
C-pos.1	"本"の上に茶点 brown dot above "本"		
C-pos.9	"大"の右下方に黒点 black dot to lower right of "大" of imprint		
D-pos.4	肩に白点 white dot on shoulder		
D-pos.5	頭の右方に茶点 brown dot to upper right of head		
D-pos.6	第1マージンに茶線 brown line in NW margin		
D-pos.9	"大"の右に黒点 black dot to right of "大" of imprint		
D-pos.10	第4マージンに茶点 brown dot in SW margin		

C388 赤十字規約制定100年記念 / Centenary of International Red Cross

発行日：1963（昭和38）.5.8
Date of Issue：1963 (Showa 38) .5.8

●特印 Special Date Stamp
東京 Tokyo

C388 世界地図と赤十字100年祭マーク
World Map and Centenary Emblem of Red Cross

		**	●	FDC
C388 10 yen 多色 multicolored		50	30	250

切手データ Basic Information	
版式：グラビア輪転版	printing plate : rotary photogravure plate
目打：櫛型13½	perforation : comb perf. 13½
印面寸法：33×22.5mm	printing area size : 33×22.5mm
発行枚数：1,200万枚	number issued : 12 million
シート構成：20枚（横5×縦4）	pane format : 20 subjects (5x4)
製版・目打組合せ：タイプ2B-2	plate and perforation configuration : type 2B-2

記念・特殊切手（額面「円位」時期、ローマ字なし）

C388 ● 定常変種 Constant Flaws

A-pos.1　B-pos.2　C-pos.7　D-pos.7

位置 position	特徴 feature	**	●
A-pos.1	右地図の上に灰緑点 gray-green dot above right map	200	100
B-pos.2	ヨットの帆に濃い点 dark dot on sail of yacht		
C-pos.7	"日"の左に灰緑点 gray-green dot to left of "日"		
D-pos.7	アフリカ大陸に大きな濃い点2個 two large dark dots on African continent		

C389　第5回国際かんがい排水委員会総会記念

発行日：1963（昭和38）.5.15
5th Plenary Conference, International Commission on Irrigation and Drainage
Date of Issue：1963 (Showa 38) .5.15

C389 地球と水路を表す木の葉
Globe and Leaf with Symbolic River System

●特印
Special Date Stamp

東京 Tokyo

		**	●	FDC
C389 10 yen こい青 blue		50	30	250

切手データ Basic Information	
版式：グラビア輪転版	printing plate : rotary photogravure plate
目打：櫛型13½	perforation : comb perf. 13½
印面寸法：22.5×27mm	printing area size : 22.5×27mm
発行枚数：1,100万枚	number issued : 11 million
シート構成：20枚（横4×縦5）	pane format : 20 subjects (4×5)
製版・目打組合せ：タイプ2B-1	plate and perforation configuration : type 2B-1

C389 ● 定常変種 Constant Flaws

A-pos.3　A-pos.15　A-pos.20　B-pos.7

B-pos.17

C-pos.6　C-pos.7

B-pos.20

C-pos.17　D-pos.13　D-pos.14　D-pos.17

位置 position	特徴 feature	**	●
A-pos.3	葉に白点2ヵ所 two white dots on leaf	200	100
A-pos.15	"日"の左に白点2個 two white dots to left of "日"		
A-pos.20	"便"の右方に太い白線 thick white line to right of "便"		
B-pos.7	"N"の上方と葉に白点 white dots above "N" and on leaf		
B-pos.17	"9"の下方に濃い点 dark dot below "9"		
B-pos.20	"際"の上方に青点、"念"の右下に白点 blue dot above "際", white dot to lower right of "念"		
C-pos.6	"日本"の下方に白点群 white dots below "日本"		
C-pos.7	葉に白抜きと青点 white spot and blue dot on leaf		
C-pos.17	日本の右に玉状の白もや、"1"の左方に白点 white haze to right of Japan, white dot to lower left of "1"		
D-pos.13	"0"の右下に点と線 dot and line to lower right of "0"		
D-pos.14	"GA"の右に濃い点 dark dot to right of "GA"		
D-pos.17	線の先端の下とマージンに濃い点 dark dots below line tip and in margin		

記念・特殊切手（額面「円位」時期、ローマ字なし）

C390-395　鳥シリーズ　発行年：1963-1964（昭和38-39）
Bard Series　Year of Issue : 1963-1964 (Showa38-39)

C390 ルリカケス
Purple Jay

C391 ライチョウ
Rock Ptarmigan

C392 キジバト
Eastern Turtle Dove

C393 コウノトリ
Oriental White Stork

C394 ウグイス
Bush Warbler

C395 ホオジロ
Meadow Bunting

● 風景印
Pictrial Postmarks

C390 名瀬
Naze

C391 松本
Matsumoto

C391 富山
Toyama

C392 熊谷
Kumagaya

C393 豊岡
Toyooka

C394 枚岡
Hiraoka

C395 佐倉
Sakura

	**	●	FDC
C390 10 yen 多色 multicolored (1963.6.10)	120	100	550
C391 10 yen 多色 multicolored (1963.8.10)	50	40	450
C392 10 yen 多色 multicolored (1963.11.20)	50	40	380
C393 10 yen 多色 multicolored (1964.1.10)	50	40	380
C394 10 yen 多色 multicolored (1964.2.10)	50	40	380
C395 10 yen 多色 multicolored (1964.5.1)	50	40	380
C390-395 (6種 Set of 6 Values)	370	300	

切手データ Basic Information

版式：グラビア輪転版	printing plate : rotary photogravure plate
目打：櫛型13½	perforation : comb perf. 13½
印面寸法：27×31mm	printing area size : 27×31mm
発行枚数：C390=1,350万枚、C391=1,690万枚、C392=1,720万枚、C393=1,750万枚、C394=1,800万枚、C395=2,500万枚	number Issued : C390=13.5 million, C391=16.9 million, C392=17.2 million, C393=17.5 million, C394=18 million, C395=25 million
シート構成：20枚（横4×縦5）	pane format : 20 subjects (4×5)
製版・目打組合せ：C390=タイプ1C-3、1C-5、C391=タイプ1C-3、C392-395=タイプ1C-5	plate and perforation configuration : C390=type 1C-3 and 1C-5, C391=type 1C-3, C392-395=type 1C-5

● トンボ Marginal Guide Marks

1) C390、C391の実用版はシート4面を田型状に配置して中央にトンボがあり、裁断後もシート四隅の耳紙にあるトンボの線によりシートA～Dの配置が確定できる。
2) C390の製版・組合せは、タイプ1C-3とタイプ1C-5の併用がシートA、Bの定常変種により確定できる。タイプ1C-5は初めて採用された。
3) C392は裁断がずれたことにより右耳紙にトンボが残っているが、この時期の切手としては珍しい。上耳紙の黄色の縦線と緑色の横線は裁断用のトンボであり、中央の水色のトンボは目打穿孔用と考えられる。

1) The plate configuration of C390 and C391 is a block of four panes with a marginal guide mark at the center of the four panes. The position of a pane can thus be identified by the marginal guide mark.
2) The plate and perforation configurations of C390 are identified as types 1C-3 and 1C-5 by constant flaws in panes A and B.
3) A pane cutting shift causes a marginal guide mark to appear on the right selvage of C392, a scarce occurrence for stamps in this period. The yellow vertical and green horizontal marginal guide marks at NE of the selvage are considered to be guides for cutting, and the light blue marginal guide mark at center of the right selvage is considered to be a guide for perforation.

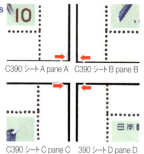
C390 シート A pane A　C390 シート B pane B
C390 シート C pane C　390 シート D pane D

C392

● 不整目打 Irregular Perforation

（左 left）C391
（右 right）C392

1) C390、C391は目打ピッチと切手縦寸が整合していないため、目打継ぎ目に顕著な寸詰まりがある。この不整目打はC392以降解消された。
1) Since the perforation pitch did not match the height of the printing area for C390 and C391, the comb perforation joints of these stamps were noticeably shortened. This perforation irregularity was resolved for C392 and thereafter.

記念・特殊切手（額面「円位」時期、ローマ字なし）

C390 ●定常変種 Constant Flaws

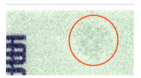
A-pos.4

A-pos.14

B-pos.4

B-pos.9　　　　　B-pos.10

C-pos.7

C-pos.16

D-pos.11

位置 position	特徴 feature	**	●
A-pos.4	"便"の右方にシミ状の濃い緑点 dark green spot to right of "便"	300	200
A-pos.14	"日"の左に白点 white dot to left of "日"		
B-pos.4	"け"の上方に白点 white dot above "け"		
B-pos.9	白点と茶点 white dot and brown dot		
B-pos.10	目の上方に茶点 brown dot above eye		
C-pos.7	首の右方に白点 white dot to right of neck		
C-pos.16	"便"の右方に白点、首に茶版白抜け white dot to right of "便", brown spot on neck.		
D-pos.11	葉に濃い緑点群 dark green dots on leaves		

C391 ●定常変種 Constant Flaws

A-pos.17

B-pos.9

B-pos.12

C-pos.7

D-pos.8

D-pos.12

位置 position	特徴 feature	**	●
A-pos.17	"ら"の左下方に大きい白点 large white dot to lower left of "ら"	200	100
B-pos.9	尾羽の左に黒点 black dot to left of tail		
B-pos.12	"う"の右方に黒点 black dot to right of "う"		
C-pos.7	腹の下に黒点 black dot below belly		
D-pos.8	"0"の下に黒点 black dot below "0"		
D-pos.12	下マージンに赤線 red line in bottom margin		

C392 ●定常変種 Constant Flaws

A-pos.16

B-pos.5

B-pos.17

B-pos.19

C-pos.20

D-pos.2

D-pos.16

位置 position	特徴 feature	**	●
A-pos.16	首の左下方に赤点 red dot to lower left of neck	200	100
B-pos.5	"本"の上に小さな赤点 small red dot above "本"		
B-pos.17	葉の間に小さな緑点 small green dot between leaves		
B-pos.19	"1"の上方に黒点 black dot above "1"		
C-pos.20	"き"の左右に小さな赤点2ヵ所 two small red dots to left of "き"		
D-pos.2	尾羽の下に小さな黒点 small black dot below tail		
D-pos.16	枝の下に小さな赤点 small red dot below branch		

C393 ●定常変種 Constant Flaws

A-pos.7

A-pos.18

B-pos.5

B-pos.6

B-pos.11

C-pos.5

C-pos.12

C-pos.16　　　D-pos.19

73

記念・特殊切手（額面「円位」時期、ローマ字なし）

位置 position	特徴 feature	**	●
A-pos.7	"日"の赤点化、首の右に白ぼやけ red dot on "日", white blur to right of neck	200	100
A-pos.18	くちばしの下方に白点 white dot below beak		
B-pos.5	首に黒点 black dot on neck		
B-pos.6	くちばしの上に白点 white dot above beak		
B-pos.11	くちばしの下方に白点 white dot below beak		
C-pos.5	"こう"の上方に白点 white dot above "こう"		
C-pos.12	尾羽の上方に赤点2ヵ所 two red dots above tail		
C-pos.16	"本"の下方に濃い点 dark dot below "本"		
D-pos.19	くちばしの右下方に白点 white dot to lower right of beak		

C394 ●定常変種 Constant Flaws

A-pos.8　　B-pos.11　　C-pos.18　　D-pos.18

位置 position	特徴 feature	**	●
A-pos.8	尾羽の上方にリタッチ状の大きな濃い緑点 large dark green retouch above tail	200	100
B-pos.11	頭の上方に大きな白ぼやけ large white blur above head		
C-pos.18	つぼみの右に大きな濃い緑点 large dark green dot to right of bud		
D-pos.18	下腹の右に白点 white dot to right of lower abdomen		

C395 ●定常変種 Constant Flaws

A-pos.6　　A-pos.7

B-pos.11

B-pos.10

C-pos.10　　C-pos.12　　D-pos.14　　D-pos.15

位置 position	特徴 feature	**	●
A-pos.6	つぼみの右に白点 white dot to right of bud	200	100
A-pos.7	"ほ"の左下方に白点 white dot to lower left of "ほ"		
B-pos.10	くちばしの上に白点 white dot above beak		
B-pos.11	"0"の上に白点 white dot above "0"		
C-pos.10	"本"の左方に白点 white dot to left of "本"		
C-pos.12	"1"の上に長い白線 long white line above "1"		
D-pos.14	つぼみの左上方に円状の白ぼやけ circular white blur to upper left of bud		
D-pos.15	"便"の左に円状の白ぼやけ circular white blur to left of "便"		

C396　名神高速道路開通記念　発行日：1963（昭和38）.7.15
Opening of Meishin（Nagoya-Kobe）Expressway　Date of Issue：1963 (Showa 38) .7.15

C396 栗東インターチェンジ
Ritto Interchange

●特印
Special Date Stamp

京都 Kyoto

	**	●	FDC
C396 10 yen 青緑、オリーブ黒、黄味だいだい blue green, black and orange	50	30	270

C396 ●定常変種 Constant Flaws

A-pos.1　　B-pos.15　　B-pos.17

C-pos.11　　C-pos.16

C-pos.12

D-pos.12

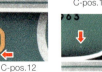

C-pos.17

切手データ Basic Information

版式：グラビア輪転版	printing plate : rotary photogravure plate
目打：櫛型13½	perforation : comb perf. 13½
印面寸法：33×22.5mm	printing area size : 33×22.5mm
発行枚数：1,300万枚	number issued : 13 million
シート構成：20枚（横5×縦4）	pane format : 20 subjects (5x4)
製版・目打組合せ：タイプ2B-2	plate and perforation configuration : type 2B-2

位置 position	特徴 feature	**	●
A-pos.1	道路の中央に黒点 black dot in center of road	200	100
B-pos.15	道路の外に白ぼやけ2ヵ所 two white blur beside road		
B-pos.17	"便"の下方に大きな白ぼやけ large white blur below "便"		
C-pos.11	道路に黒点 black dot on road		
C-pos.12	"3"の右方に白点、"0"の下に白点 white dot to right of "3", white dot below "0"		
C-pos.16	道路に小さな黒点 small black dot on road		
C-pos.17	"本"の下に白点、"通"の下に緑点 white dot below "本", green dot below "通"		
D-pos.12	道路の外に白ぼやけ white blur beside road		

C397 ガールスカウト・アジア大会記念 / Asian Girl Scout Camp

発行日：1963（昭和38）.8.1　Date of Issue : 1963 (Showa 38).8.1

C397 ガールスカウトと世界連盟旗 Girl Scout and Flag

●特印 Special Date Stamp

戸隠 Togakushi

		**	●	FDC
C397	10 yen 多色 multicolored	50	30	250

切手データ Basic Information

版式：グラビア輪転版	printing plate : rotary photogravure plate
目打：櫛型13½	perforation : comb perf. 13½
印面寸法：22.5×27mm	printing area size : 22.5×27mm
発行枚数：1,350万枚	number issued : 13.5 million
シート構成：20枚（横4×縦5）	pane format : 20 subjects (4×5)
製版・目打組合せ：タイプ2B-1	plate and perforation configuration : type 2B-1

1）紫色のスクリーン角度には83度と86度が確認されているが、判別は難しい。
1) Two screen angles of 83° and 86° are confirmed in the purple plate, but these are difficult to identify.

C397 ● 定常変種 Constant Flaws

A-pos.16　B-pos.15　B-pos.18　D-pos.2

位置 position	特徴 feature	**	●
A-pos.16	肩の上に赤点 red dot above shoulder	200	100
B-pos.15	ポールの右上方に紫線 purple line to upper right of pole		
B-pos.18	ポールの左上方に青点 blue dot to upper left of pole		
D-pos.2	"本"の左下方に白ぼやけ white blur to lower left of "本"		

C398 国際電波科学連合第14回総会記念 / 14th Conference, International Union of Radio Science

発行日：1963（昭和38）.9.9　Date of Issue : 1963 (Showa 38).9.9

C398 パラボラアンテナ Parabolic Antenna

●特印 Special Date Stamp

東京 Tokyo

		**	●	FDC
C398	10 yen 多色 multicolored	50	30	250

切手データ Basic Information

版式：グラビア輪転版	printing plate : rotary photogravure plate
目打：櫛型13½	perforation : comb perf. 13½
印面寸法：22.5×27mm	printing area size : 22.5×27mm
発行枚数：1,410万枚	number issued : 14.1 million
シート構成：20枚（横4×縦5）	pane format : 20 subjects (4×5)
製版・目打組合せ：タイプ2B-1	plate and perforation configuration : type 2B-1

C398 ● 定常変種 Constant Flaws

A-pos.19　B-pos.5　C-pos.1

C-pos.11　C-pos.13　D-pos.1

位置 position	特徴 feature	**	●
A-pos.19	アンテナ先端の左に小さな青点 small blue dot to left of antenna tip	200	100
B-pos.5	シミ状の大きな濃い点 large dark stain-like dot		
C-pos.1	"O"と"U"の間の左に青点 blue dot to left of "O" and "U"		
C-pos.11	"日"の左下方に小さな青点 small blue dot to lower left of "日"		
C-pos.13	"E"の右に青点 blue dot to right of "E"		
D-pos.1	柱の中央に小さな青点 small blue dot on center of pilla		

記念・特殊切手（額面「円位」時期、ローマ字なし）

C399 国際文通週間 発行日：1963（昭和38）.10.6
International Letter Writing Week 1963   Date of Issue : 1963 (Showa 38) .10.6

C399 葛飾北斎画「富嶽三十六景 神奈川沖浪裏」
"Thirty-six Views of Mt. Fuji : The Great Wave Off the Coast of Kanagawa" by Katsushika Hokusai

●特印 Special Date Stamp

横浜 Yokohama

	**	●	FDC
C399 40 yen 多色 multicolored	700	350	2,000

切手データ Basic Information	
版式：グラビア輪転版	printing plate : rotary photogravure plate
目打：櫛型13	perforation : comb perf. 13
印面寸法：35.5×25mm	printing area size : 33.5×25mm
発行枚数：750万枚	number issued : 7.5 million
シート構成：10枚（横5×縦2）	pane format : 10 subjects (5x2)
製版・目打組合せ：タイプ1F	plate and perforation configuration : type 1F

C399 ●定常変種 Constant Flaws

A pos.1

C-pos.4/5

C-pos.9

B-pos.2

B-pos.8

A-pos.10

D-pos.4

位置 position	特徴 feature	**	●
A-pos.1	郵に青の破線 blue dashed line on "郵"	2,000	1,000
A-pos.10	大波の右に大きな黒点 large black dot to right of big wave		
B-pos.2	舟の上方に白抜け white spot above boat		
B-pos.8	"4"の中に青点 blue dot on "4"		
C-pos.4/5	"K"の右に青点、"通"の右に青点 blue dot to right of "K", blue dot to right of "通"		
C-pos.9	波に青点2個 two blue dots in waves		
D-pos.4	波に青点 blue dot in waves		

1) 製版・目打組合せタイプ1Fの最初の切手である。
1) C399 is the first stamp with plate and configuration type 1F.

C400 東京国際スポーツ大会記念 発行日：1963（昭和38）.10.11
Tokyo International Sports Meet (Pre-Olympic)   Date of Issue : 1963 (Showa 38) .10.11

C400 飛び込み、棒高跳び、短距離走
Diving, Pole Vault and Sprint

●特印 Special Date Stamp

東京 Tokyo

	**	●	FDC
C400 10 yen 多色 multicolored	50	30	250

切手データ Basic Information	
版式：グラビア輪転版	printing plate : rotary photogravure plate
目打：櫛型13½	perforation : comb perf. 13½
印面寸法：22.5×33mm	printing area size : 22.5×33mm
発行枚数：1,500万枚	number issued : 15 million
シート構成：20枚（横4×縦5）	pane format : 20 subjects (4x5)
製版・目打組合せ：タイプ2B-1	plate and perforation configuration : type 2B-1

C400 ●定常変種 Constant Flaws

A-pos.5

A-pos.13

A-pos.17

A-pos.19

C-pos.3

C-pos.20

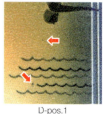
D-pos.1

B-pos.1

B-pos.10

B-pos.13

B-pos.16

記念・特殊切手（額面「円位」時期、ローマ字なし）

位置 position	特徴 feature	**	●
A-pos.5	"便"の右方に小さな黒点 small black dot to right of "便"	200	100
A-pos.13	"1"の左上方に小さな黒点 small black dot to upper left of "1"		
A-pos.17	"京"と"国"の間に黒点 black dot between "京" and "国"		
A-pos.19	肘の左上方に黒点 black dot to upper left of elbow		
B-pos.1	"1"の左に黒点 black dot to left of "1"		
B-pos.10	波の下方に小さな青点群 small blue dots below wave		
B-pos.13	"便"の右方に小さな黒点 small black dot to right of "便"		
B-pos.16	脚の右に小さな青点2ヵ所 two small blue dots to right of leg		
C-pos.3	ポールに青傷 blue scratch on pole		
C-pos.20	"日"の左下と"便"の上に黒点 black dots to lower left of "日" and above "便"		
D-pos.1	波に青点、頭の下方に小さな青点 blue dot to left of wave, small blue dot below head		

C401-402　第18回国体記念
18th National Athletic Meet

発行日：1963（昭和38）.10.27
Date of Issue：1963（Showa 38）.10.27

	**	●	FDC
C401 5 yen 灰味茶 brown	40	30	
C402 5 yen 暗い緑 slate green	40	30	
C322-323（2種連刷 pair）	90	200	250

 C401 相撲 Sumo
 C402 徒手体操 Calisthenics

● 特印 Special Date Stamp

山口 Yamaguchi

切手データ Basic Information	
版式：ザンメル凹版	printing plate : sammeldruck intaglio printing
目打：櫛型13½	perforation : comb perf. 13½
印面寸法：27×22.5mm	printing area size : 27×22.5mm
発行枚数：各1,500万枚	number issued : 15 million each
シート構成：20枚（横5×縦4）	pane format : 20 subjects (5x4)
製版・目打組合せ：タイプ1C-2	plate and perforation configuration : type 1C-2

C403-406　お祭りシリーズ
Festival Series

発行年：1964-1965（昭和39-40）
Year of Issue：1964-1965（Showa39-40）

 C403 高山祭 Takayama Festival
 C404 祇園祭 Gion Festival
 C405 相馬野馬追 Souma Nomaoi
 C406 秩父夜祭 Chichibu Night Festival

切手データ Basic Information	
版式：グラビア輪転版	printing plate : rotary photogravure plate
目打：C403-404＝櫛型13½×13、 C405-406＝櫛型13×13½	perforation : C403-404=comb perf. 13½ ×13, C405-406=comb perf. 13×13½
印面寸法：C403-404＝22.5×40 mm、C405-406＝40×22.5mm	printing area size : C403-404=22.5×27mm, C405-406=40×22.5mm
発行枚数：C403＝2,300万枚、C404＝ 2,800万枚、C405-406＝各3,000万枚	number issued : C403=23 million, C404=28 million, C405-406=30 million each
シート構成：C403-404＝20枚（横4× 縦5）、C405-406＝20枚（横5×縦4）	pane format : C403-404=20 subjects (4x5), C405-406=20 subjects (5x4)
製版・目打組合せ：C403-404＝タイ プ2B-1、C405-406＝タイプ2B-2	plate and perforation configuration : C403- 404=type 2B-1, C405-406=type 2B-2

● 風景印 Pictrial Postmarks

 C403 高山 Takayama
 C404 京都祇園 Kyoto Gion
 C405 原町 Haramachi
 C406 秩父 Chichibu

	**	●	FDC
C403 10 yen 多色 multicolored (1964.4.15)	50	50	400
C404 10 yen 多色 multicolored (1964.7.15)	50	30	400
C405 10 yen 多色 multicolored (1965.7.16)	50	30	350
C406 10 yen 多色 multicolored (1965.12.3)	50	30	400
C403-406（4種 Set of 4 Values）	200	140	

● 連続櫛型目打二連型 Continuous Comb Perforation, Double Comb

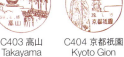

1) C403、C404の製版・目打組合せは上下の耳紙に目打穴が飛び出さない二連型であり、この2種だけに使用された。

1) The perforations of C403 and C404 are the double comb perforation type without extensions into the top and bottom selvages. This configuration was used for only C403 and C404.

記念・特殊切手（額面「円位」時期、ローマ字なし）

C403 ●定常変種 Constant Flaws

A-pos.14

C-pos.13

C-pos.20

B-pos.2

D-pos.5

D-pos.8

D-pos.13

D-pos.16

位置 position	特徴 feature	**	●
A-pos.14	山鉾の左に白点 white dot to left of float	200	100
B-pos.2	"日"の左下方に赤点 red dot to lower left of "日"		
C-pos.13	山鉾の屋根の左方に白点 white dot to left of float		
C-pos.20	山鉾の手摺の右に白点 white dot to right of float		
D-pos.5	"0"の中に黒点 black dot in "0"		
D-pos.8	"日"の上方に黒の破線 black dashed line above "日"		
D-pos.13	山鉾の屋根の左に白点 white dot to left of float		
D-pos.16	"便"の左上方に白点 white dot to upper left of "便"		

C404 ●定常変種 Constant Flaws

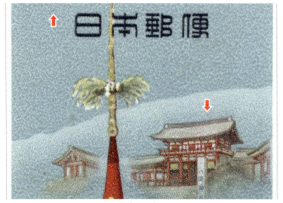
A-pos.5

B-pos.3

B-pos.9

B-pos.18

D-pos.6

D-pos.18

位置 position	特徴 feature	**	●
A-pos.5	"日"の左上方と屋根の上に赤点 red dots to upper left of "日" and above roof	200	100
B-pos.3	左マージンに赤点 red dot in left margin		
B-pos.9	"祇"の文字欠け defect in "祇"		
B-pos.18	"0"の中に灰色の突起 gray protrusion in "0"		
D-pos.6	"祇"の左下方に白線 white line to lower left of "祇"		
D-pos.18	山鉾の左に白点 white dot to left of float		

●スクリーン線数 Screen Line Number

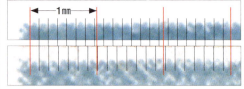

260線（網点22個／3mm）260 lines
250線（網点21個／3mm）250 lines

1) C404の青色のスクリーン線数は、260線と250線の2種があるが、識別は困難である。
1) There are two types of blue screen line number for C404: 260 lines and 250 lines, it is difficult to distinguish them.

C405 ●定常変種 Constant Flaws

A-pos.12　A-pos.17

B-pos.20

B-pos.13　C-pos.5

記念・特殊切手（額面「円位」時期、ローマ字なし）

C-pos.7

D-pos.9

D-pos.5

C406 ● 定常変種 Constant Flaws

A-pos.5

A-pos.20

B-pos.9

B-pos.14

B-pos.19　B-pos.7

D-pos.20

D-pos.10

位置 position	特徴 feature	**	●
A-pos.12	"追"の右下方に赤点 red dot to lower right of "追"	200	100
A-pos.17	"0"の中に小さな赤点 small red dot in "0"		
B-pos.13	右マージンに緑点 green dot in right margin		
B-pos.20	旗に青点 blue dot on flag		
C-pos.5	"郵"の右上方に青の突起 blue protrusion to upper right of "郵"		
C-pos.7	"追"の上方に赤点2ヵ所 two red dots above "追"		
D-pos.5	"追"の右方に赤点 red dot to right of "追"		
D-pos.9	"0"の下に大きな赤点 large red dot below "0"		

位置 position	特徴 feature	**	●
A-pos.5	第2コーナーのマージンに青点 blue dot in margin of NE corner	200	100
A-pos.20	第3コーナーのマージンに赤点 red dot in margin of SE corner		
B-pos.9	山鉾の屋根の左に赤点 red dot to left of float		
B-pos.14	下マージンに小さな赤点 small red dot in bottom margin		
B-pos.19	"大"の左上方に黒点（銘版）black dot to upper left of "大" of imprint		
D-pos.7	上マージンに茶点 brown dot in top margin		
D-pos.10	"0"の右上方に赤点 red dot to upper right of "0"		
D-pos.20	"日"の中に小さな白点 small white dot in "日"		

C407 切手趣味週間 Philatelic Week 1964
発行日：1964（昭和39）.4.20
Date of Issue：1964 (Showa 39) .4.20

C407「源氏物語絵巻：宿木 三」
"The Tale of Genji Illustrated Scrolls: Yadorigi (Ivy) 3"

● 特印 Special Date Stamp

東京 Tokyo

	**	●	FDC
C407 10 yen 多色 multicolored	60	50	450

切手データ Basic Information	
版式：グラビア輪転版	printing plate : rotary photogravure plate
目打：櫛型13½	perforation : comb perf. 13½
印面寸法：48×33mm	printing area size : 48×33mm
発行枚数：2,800万枚	number issued : 28 million
シート構成：10枚（横2×縦5）	pane format : 10 subjects (2×5)
製版・目打組合せ：タイプ1C-3	plate and perforation configuration : type 1C-3

● 刷色とスクリーン角度 Printing Color and Screen Angle

タイプ type	赤紫 magenta		黄茶 yellow-brown			黒 black				緑 green	
	スクリーン角度 screen angle										
	58	87	67	72	75	3	85	90①	90②	45	90
1	●		●				●			●	
2	●		●				●			未確定 unconfirmed	
3	●			●		●					
4	●			●				●			
5					●				●		

1) 刷色の4色すべてにスクリーン角度が複数あり、刷色とスクリーン角度の組合せで実用版をタイプ1～5に分類した。タイプ5が最も少なく、各色のスクリーン角度別に主な定常変種を掲載している。

1) All four printing colors have multiple screen angles. The plates are classified into types 1 - 5 according to the combination of screen angle and color. Among them, type 5 is the scarcest. The constant flaws are described for each screen angle and color combination.

記念・特殊切手（額面「円位」時期、ローマ字なし）

C407 ●刷色・スクリーン角度別定常変種 Constant Flaws by Printing Color and Screen Angle

※ 共通の特徴 Common features

赤紫58度 magenta 58 degrees

A-pos.8

C-pos.1

C-pos.4

D-pos.8

B-pos.4

赤紫87度 magenta 87 degrees　　黄茶75度 yellow-brown 75 degrees

B-pos.4　　A-pos.8

黒3度 black 3 degrees　　黒90度① black 90 degrees ①　　黒90度② black 90 degrees ②

D-pos.8

A-pos.4

C-pos.3

位置 position	特徴 feature	**	●
赤紫58度 magenta 58 degrees		200	100
A-pos.8	左の柱に赤点 red dot on left pillar		
B-pos.4※	左右の柱に赤点 red dots on left and right pillars		
C-pos.1	左マージンに赤点 red dot in left margin		
C-pos.4	右マージンに赤破線 dashed red line in right margin		
D-pos.8	床に大きな丸いシミ large round stain on floor		
赤紫87度 magenta 87 degrees			
B-pos.4※	左右の柱に赤点 red dots on left and right pillars		
黄茶75度 yellow-brown 75 degrees			
A-pos.8	床に半円状の茶線 brown arc on floor		
黒3度 black 3 degrees			
D-pos.8	"本"に黒点 black dot in "本"		
黒90度① black 90 degrees ①			
A-pos.4	女御の頭の上に黒点 black dot avobe woman's head		
黒90度② black 90 degrees ②			
C-pos.3	左マージンに黒点 black dot in left margin		

1）黒90度①は89度に近い角度、黒90度②は90度である。
1) "Black 90° ①" is close to 89°, whereas "Black 90° ②" is exactly 90°.

C408　姫路城修理完成記念　Restoration of Himeji Castle
発行日：1964（昭和39）.6.1　Date of Issue：1964 (Showa 39) .6.1

C408 姫路城 Himeji Castle

●特印 Special Date Stamp

姫路 Himeji

		**	●	FDC
C408 10 yen 暗い茶 dark brown		50	30	200

切手データ Basic Information	
版式：グラビア輪転版	printing plate : rotary photogravure plate
目打：櫛型13½	perforation : comb perf. 13½
印面寸法：27×22.5mm	printing area size : 27×22.5mm
発行枚数：2,200万枚	number issued : 22 million
シート構成：20枚（横5×縦4）	pane format : 20 subjects (5x4)
製版・目打組合せ：タイプ2B-2	plate and perforation configuration : type 2B-2

C408 ●定常変種 Constant Flaws

A-pos.2　A-pos.11　B-pos.5　　B-pos.8

B-pos.13　C-pos.15　D-pos.3　D-pos.20

位置 position	特徴 feature	**	●
A-pos.2	"1964"の下に円弧状の線 arc below "1964"	200	100
A-pos.11	山肌に茶点 brown dot on mountain		
B-pos.5	天守閣の右に濃い点 dark dot to right of castle		
B-pos.8	天守閣の左に濃い点 dark dot to left of castle		
B-pos.13	"日"の下に茶点 brown dot below "日"		
C-pos.15	城郭の上に濃い点 dark spot above castle wall		
D-pos.3	天守閣の右に茶点 brown dot to right of castle		
D-pos.20	"1"の左下方に玉状のシミ bead-shaped stain to lower left of "1"		

C409-410 第19回国体記念
19th National Athletic Meet

発行日：1964（昭和39）.6.6　Date of Issue：1964 (Showa 39) .6.6

C409 ハンドボール Handball　　C410 平均台 Balance Beam

●特印 Special Date Stamp

新潟 Niigata

	**	●	FDC
C409 5 yen 青味灰 bluish black	40	30	
C410 5 yen 赤 red	40	30	
C409-410（2種連刷 pair）	90	100	200

切手データ Basic Information

版式： ザンメル凹版	printing plate : sammeldruck intaglio printing
目打：櫛型13½	perforation : comb perf. 13½
印刷寸法：22.5×27mm	printing area size : 22.5×27mm
発行枚数：各2,200万枚	number issued : 22 million each
シート構成：20枚（横4×縦5）	pane format : 20 subjects (4x5)
製版・目打組合せ：タイプ1C-5	plate and perforation configuration : type 1C-5

C411 太平洋横断ケーブル開通記念
Opening of Transpacific Submarine Cable

発行日：1964（昭和39）.6.19　Date of Issue：1964 (Showa 39) .6.19

C411 ケーブル断面と経由地点 Cable Cross Section and Route Points

●特印 Special Date Stamp

二宮 Ninomiya

	**	●	FDC
C411 10 yen くすみ黄だいだい、緑味灰、こい赤紫 yellow, gray green and deep magenta	50	30	200

切手データ Basic Information

版式：グラビア輪転版	printing plate : rotary photogravure plate
目打：櫛型13½	perforation : comb perf. 13½
印刷寸法：33×22.5mm	printing area size : 33×22.5mm
発行枚数：2,400万枚	number issued : 24 million
シート構成：20枚（横5×縦4）	pane format : 20 subjects (5x4)
製版・目打組合せ：タイプ2B-2	plate and perforation configuration : type 2B-2

●スクリーン角度 Screen Angles

 1版 Plate 1　赤紫23度 magenta 23 degree

 2版 Plate 2　赤紫75度 magenta 75 degree

1) 赤紫色の実用版には、スクリーン角度が23度と75度の2種類がある。23度の実用版を含むユニットで印刷された方を1版、75度を含むユニットで印刷された方を2版に分類した。

1) There are two screen angles for the magenta plate: 23 degrees and 75 degrees. The plate printed from units including the 23-degree magenta plate is classified as plate 1, and that including the 75-degree magenta plate is classified as plate 2.

C411 ●定常変種 Constant Flaws

 P1/P2-A-pos.1

 P1/P2-A-pos.6

 P2-A-pos.5

 P1/P2-B-pos.1

 P2-B-pos.11

 P1/P2-C-pos.3

 P1/P2-C-pos.6

 P2-C-pos.20

 P1-D-pos.15

 P2-D-pos.4

 P2-D-pos.7

位置 position	特徴 feature	**	●
P1/P2-A-pos.1	"日"の左右に大きな白点 large white dot to left of "日"	200	100
P1/P2-A-pos.6	半島状の白抜け peninsula-shaped white spot		
P2-A-pos.5	第2マージンに赤紫点 magenta dot in NE margin		
P1/P2-B-pos.1	"1"の左に白点 white dot to left of "1"		
P1/P2-B-pos.14	"1"の右下に白抜け white spot to lower right of "1"		
P2-B-pos.11	"開"の上方に白抜け斜線 white diagonal line above "開"		
P1/P2-C-pos.3	"64"の右上方に長い斜線 long diagonal line to upper right of "64"		
P1/P2-C-pos.6	"太"の左上方に大きな白点 large white dot to upper left of "太"		
P2-C-pos.20	ケーブル断面の左下に白点 white dot to lower left of cable cross section		
P1-D-pos.15	第1マージンに赤紫点2個 two magenta dots in NW margin		
P2-D-pos.4	第2コーナーに大きな白点 large white dot in NE corner		
P2-D-pos.7	"便"の右下方に白抜け white spot to lower right of "便"		

C412 首都高速道路開通記念
Opening of Metropolitan Expressway
発行日：1964（昭和39）.8.1　Date of Issue：1964 (Showa 39).8.1

C412 首都高速道路（日本橋） Metropolitan Expressway (Nihonbashi)

●特印 Special Date Stamp

東京 Tokyo

C412 10 yen くすみ緑、暗いオリーブ緑、銀 green, black and silver ………… ** 50　● 30　FDC 200

B-pos.9	B-pos.10
C-pos.3	
C-pos.4	D-pos.3 D-pos.11

切手データ Basic Information

版式：グラビア輪転版	printing plate : rotary photogravure plate
目打：櫛型13½	perforation : comb perf. 13½
印面寸法：22.5×27mm	printing area size : 22.5×27mm
発行枚数：2,400万枚	number issued : 24 million
シート構成：20枚（横4×縦5）	pane format : 20 subjects (4×5)
製版・目打組合せ：タイプ2B-1	plate and perforation configuration : type 2B-1

1) この切手以降、グラビア印刷の記念・特殊切手のスクリーン線数は250線となった。
1) The photogravure screen became 250 lines for this stamp and thereafter.

C412 ●定常変種 Constant Flaws

 A-pos.5　 A-pos.11

位置 position	特徴 feature	**	●
A-pos.5	後ろの欄干に大きな白抜け large white spot in rear railing	200	100
A-pos.11	下マージンに小さな緑点2ヵ所 two small green dots in bottom margin		
B-pos.9	"郵"の右下方に白抜け white spot to lower right of "郵"		
B-pos.10	前の欄干に大きな緑点と大きな白抜け large green dot and large white spot in front railing		
C-pos.3	前の欄干に白抜け white spot in front railing		
C-pos.4	第1コーナーと第3コーナーのマージンに緑点 green dots in NW and SE margins		
D-pos.3	前の欄干に白抜けと緑点 white spot and green dot in front railing		
D-pos.11	前の欄干に大きなシミ large stain in front railing		

C413 国際通貨基金・国際復興開発銀行東京総会記念
International Monetary Fund (IMF) and International Bank for Reconstruction and Development (IBRD) Tokyo General Meeting
発行日：1964（昭和39）.9.7　Date of Issue：1964 (Showa 39).9.7

C413 各機関の頭文字の硬貨 Coins with Organizations' Initials

●特印 Special Date Stamp

東京 Tokyo

C413 10 yen 多色 multicolored ………… ** 50　● 30　FDC 200

切手データ Basic Information

版式：グラビア輪転版	printing plate : rotary photogravure plate
目打：櫛型13½	perforation : comb perf. 13½
印面寸法：33×22.5mm	printing area size : 33×22.5mm
発行枚数：2,400万枚	number issued : 24 million
シート構成：20枚（横5×縦4）	pane format : 20 subjects (5×4)
製版・目打組合せ：タイプ2B-2	plate and perforation configuration : type 2B-2

C413 ●定常変種 Constant Flaws

 A-pos.8

 A-pos.14

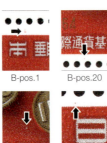 B-pos.1　B-pos.20

 C-pos.16　C-pos.19

 D-pos.14

位置 position	特徴 feature	**	●
A-pos.8	下マージンに赤点 red dot in bottom margin	200	100
A-pos.14	硬貨の上方に大きな白点、"際"の下に白点、下マージンに小さな赤点 large white dot above coin, white dot below "際", small red dot in bottom margin		
B-pos.1	上マージンに赤点3個 three red dots in top margin		
B-pos.20	下マージンに小さな赤点 small red dot in bottom margin		
C-pos.16	"国"の右上方に白抜け white spot to upper right of "国"		
C-pos.19	上マージンに小さな赤点 small red dot in top margin		
D-pos.14	上マージンに赤点2個 two red dots in top margin		

記念・特殊切手（額面「円位」時期、ローマ字なし）

C414-418　東京1964 オリンピック競技大会　発行年：1964（昭和39）
Tokyo 1964 Olympic Games　Year of Issue : 1964 (Showa 39)

C414
聖火台と競技人物
Olympic Flame and Athletes

C415　国立競技場
National Stadium

C416　日本武道館
Nippon Budokan

 (C417・C418)
C417　国立代々木競技場
Yoyogi National Gymnasium
C418　駒沢体育館
Komazawa Gymnasium

●特印 Special Date Stamp

C414 宮崎 Miyazaki　C415-418 東京 Tokyo

	**	●	FDC
C414 5 yen 多色 multicolored (1964.9.9)	30	30	300
C415 10 yen 多色 multicolored (1964.10.10)	50	30	
C416 30 yen 多色 multicolored (1964.10.10)	100	40	
C417 40 yen 多色 multicolored (1964.10.10)	120	40	
C418 50 yen 多色 multicolored (1964.10.10)	140	40	
C415-418 (4種 Set of 4 Values)		1,100	
C414-418 (5種 Set of 5 Values)	440	180	

切手データ Basic Information

版式：グラビア輪転版	printing plate : rotary photogravure plate
目打：櫛型13½	perforation : comb perf. 13½
印面寸法：C414=22.5×33mm、C415-418=33×22.5mm	printing area size : C414=22.5×33mm, C415-418=33×22.5mm
発行枚数：C414=5,500万枚、C415=4,500万枚、C416-418=各2,000万枚	number issued : C414=55 million, C415=45 million, C416-418=20 million each
シート構成：C414=20枚（横4×縦5）、C415-418=各20枚（横5×縦4）	pane format : C414=20 subjects (4×5), C415-418=20 subjects (5×4)
製版・目打組合せ：C414=タイプ2B-1、C415-418=タイプ2B-2	plate and perforation configuration : C414=type 2B-1, C415-418=type 2B-2

C414 ● スクリーン角度 Screen Angles

 1版 Plate 1 赤54度 magenta 54 degrees
 2版 Plate 2 赤66度 magenta 66 degrees
 C419 小型シート Souvenir Sheet 赤65度 magenta 65 degrees

1) スクリーン角度に赤63度が存在するとされているが、未確認である。なお、C419の5円切手のスクリーン角度は赤65度である。

1) A 63° screen angle in red has also been reported, but has not been confirmed. The red screen angle for the 5 yen stamp of C419 is 65°

● 定常変種 Constant Flaws

P1-A-pos.11

P2-A-pos.12

P1/P2-D-pos.15

(中央画像) P1-B-pos.7

P1-B-pos.14

P2-B-pos.7

P1-C-pos.2

P2-C-pos.3

P1-D-pos.6

位置 position	特徴 feature	**	●
P1-A-pos.11	下マージンに黒点2個 two black dots in bottom margin	150	100
P2-A-pos.12	"大"の下に青点2個 two blue dots below "大"		
P1-B-pos.7	"1964"の上方に黒いシミ2個、左マージンに黒点3個 two black stains above "1964", three black dots in left margin		
P1-B-pos.14	"4"の上に黒点 black dot to upper of "4"		
P2-B-pos.7	"便"の右に白抜け white spot to right of "便"		
P1-C-pos.2	聖火に黒点2個 two black dots in Olympic Flame		
P2-C-pos.3	聖火に赤点と黒点 red dot and black dot in Olympic Flame		
P1/P2-D-pos.15	上マージンに青点 blue dot in top margin		
P1-D-pos.6	右マージンに大きい黒点 large black dot in right margin		

C415 ● 定常変種 Constant Flaws

 P1-A-pos.1
P1-A-pos.8
P1-B-pos.1
 P1-C-pos.7

P1-C-pos.12
P1-D-pos.6
P2-A-pos.20
P2-B-pos.10

位置 position	特徴 feature	**	●
P1-A-pos.1	左マージンに茶点 brown dot in left margin	200	100
P1-A-pos.8	"第"の左上方に白抜け white spot to upper left of "第"		
P1-B-pos.1	右スタンドに黒点 black dot on right side seats		
P1-C-pos.7	右トラックに大きい金色点 large gold dot on right side track		

次ページにつづく Continued to next page

記念・特殊切手（額面「円位」時期、ローマ字なし）

位置 position	特徴 feature	**	●
P1-C-pos.12	左スタンドに黒点 black dot in left side seats	200	100
P1-D-pos.6	"1"の左に太い白線 thick white line to left of "1"		
P2-A-pos.20	"9"の下方に大きい濃い茶点 large dark brown dot below "9"		
P2-B-pos.10	"0"の上方に大きい濃い茶点 large dark brown dot above "0"		

1) 各色とも異なるスクリーン角度は確認されていない。現時点でシート6面分が確認されているので、便宜的に1版と2版に分類した。
1) Multiple photogravure screen angles are not known for any color. Six different panes have been identified at this point and the number of panes in one plate is four, and are classified into plate 1 and plate 2 for convenience.

C417 ●定常変種 Constant Flaws

A-pos.1　　B-pos.15　　C-pos.12　　D-pos.9

位置 position	特徴 feature	**	●
A-pos.1	屋根に大きな白抜け large white spot on roof	300	100
B-pos.15	屋根に黒点 black dot on roof		
C-pos.12	屋根に小さな赤点2個 two small red dots on roof		
D-pos.9	屋根に赤点 red dot on roof		

C416 ●定常変種 Constant Flaws

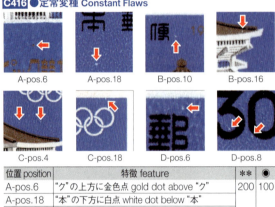

A-pos.6　A-pos.18　B-pos.10　B-pos.16
C-pos.4　C-pos.18　D-pos.6　D-pos.8

位置 position	特徴 feature	**	●
A-pos.6	"ク"の上方に金色点 gold dot above "ク"	200	100
A-pos.18	"本"の下方に白点 white dot below "本"		
B-pos.10	"便"の右下方に白点 white dot to lower right of "便"		
B-pos.16	屋根の右端に白点 white dot on right edge of roof		
C-pos.4	屋根の右上方に白点2個 two white dots to upper right of roof		
C-pos.18	五輪マークの上方に白点 white dot above Olympic mark		
D-pos.6	"郵"の上方に白点 white dot above "郵"		
D-pos.8	"30"に白点2個 two white dots around "30"		

C418 ●スクリーン角度 Screen Angles

1版 Plate 1　黒 41 度　black 41 degrees

中央の線が破線 The corner line is discontinuous.

2版 Plate 2　黒 45 度　black 45 degrees
中央の線が直線 The corner line is continuous.

●定常変種 Constant Flaws

P1-A-pos.13　P1-B-pos.10　P2-A-pos.20

P2-B-pos.18

位置 position	特徴 feature	**	●
P1-A-pos.13	"本"の上に黒点 black dot above "本"	300	100
P1-B-pos.10	"5"の左に白点 white dot to left of "5"		
P2-A-pos.20	壁に白点2ヵ所 two white dots on wall		
P2-B-pos.18	モニュメントに多数の小さな黒点 many small black dots on monument		
P2-D-pos.1	柱に大きな黒点 large black dot on pillar		

P2-D-pos.1

C419　東京1964 オリンピック競技大会 小型シート
Tokyo 1964 Olympic Games, Souvenir Sheet

発行日：1964（昭和39）.10.10
Date of Issue : 1964 (Showa 39).10.10

C419 小型シート Souvenir Sheet

タトウ paper holder

タイプA type A
当初印刷 1st printing,
文字が太い thick text

タイプB type B
追加印刷 additional printing,
文字が細い thin text

	**	●	FDC
C419 売価140円 selling price 140 yen (タトウA、B共通 type A and B common)	900	1,200	3,000

切手データ Basic Information	
版式：グラビア輪転版	printing plate : rotary photogravure plate
目打：全型13½	perforation : harrow perf. 13½
シート寸法：93×144mm	sheet size : 93×144mm
発行枚数：400万枚	number issued : 4 million

記念・特殊切手（額面「円位」時期、ローマ字なし）

C420 八郎潟干陸式記念　発行日：1964（昭和39）.9.15
Reclamation of Lake Hachirogata　Date of Issue：1964 (Showa 39).9.15

C420 稲、麦、リンゴ、乳牛
Rice, Wheat, Apple and Dairy Cow

●特印
Special Date Stamp

船越 Funakoshi

	**	●	FDC
C420 10 yen 茶紫・金 violet brown and gold	50	30	200

切手データ Basic Information	
版式：グラビア輪転版	printing plate : rotary photogravure plate
目打：櫛型13½	perforation : comb perf. 13½
印面寸法：22.5×33mm	printing area size : 22.5×33mm
発行枚数：2,400万枚	number issued : 24 million
シート構成：20枚（横4×縦5）	pane format : 20 subjects (4x5)
製版・目打組合せ：タイプ2B-1	plate and perforation configuration : type 2B-1

C420 ●定常変種 Constant Flaws

P1-A-pos.7　P1-A-pos.11　P1-B-pos.1　P1-B-pos.12　P1-C-pos.1　P1-C-pos.15　P1-D-pos.1

P1-D-pos.12　P2-A-pos.7　P2-A-pos.9　P2-B-pos.12　P2-C-pos.1

P2-D-pos.15

位置 position	特徴 feature	**	●
P1-A-pos.7	牛の右上方に茶点と白点 brown dot and white dot to upper right of cow	200	100
P1-A-pos.11	リンゴに白きず white scratch on apple		
P1-B-pos.1	穂と穂の間に茶点群 brown dots between ears of wheat		
P1-B-pos.12	りんごにリタッチ retouch on apple		
P1-C-pos.1	白点4個 four white dots		
P1-C-pos.15	手の下に白点、"1"の下方に濃い茶点 white dot below hand, dark brown dot below "1"		
P1-D-pos.1	"日"の下に大きな白点 large white dot below "日"		
P1-D-pos.12	白抜けと茶点と白点 white spots, brown dot and white dot		
P2-A-pos.7	牛の右上方に茶点 brown dot to upper right of cow		
P2-A-pos.9	"日"の右に白点 white dot to right of "日"		
P2-B-pos.12	りんごにリタッチ横長 horizontally spread retouch on apple		
P2-C-pos.1	大きな茶点と小さな茶点 large brown dot and small brown dot		
P2-D-pos.15	"0"の右上に濃い茶点、牛の尻に白点 dark brown dot to upper right of "0", white dot on hip of cow		

1) 定常変種の現れ方から実用版の同じ位置の窓口シートが2タイプ確認され、1版と2版に分類した。
1) Two types of panes printed from the same position on the plate are known, classified as "plate 1" and "plate 2".

C421 東海道新幹線開通記念　発行日：1964（昭和39）.10.1
Opening of Tokaido Shinkansen　Date of Issue：1964 (Showa 39).10.1

C421 新幹線0系特急電車
Shinkansen 0-Series

●特印
Special Date Stamp

東京 Tokyo

	**	●	FDC
C421 10 yen 青・黒 blue and black	60	30	350

切手データ Basic Information	
版式：グラビア輪転版	printing plate : rotary photogravure plate
目打：櫛型13½	perforation : comb perf. 13½
印面寸法：33×22.5mm	printing area size : 33×22.5mm
発行枚数：2,500万枚	number issued : 25 million
シート構成：20枚（横5×縦4）	pane format : 20 subjects (5x4)
製版・目打組合せ：タイプ2B-2	plate and perforation configuration : type 2B-2

記念・特殊切手（額面「円位」時期、ローマ字なし）

● スクリーン角度 Screen Angles

 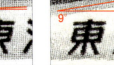

1版 Plate 1　黒2度　　2版 Plate 2　黒9度　　3版 Plate 3　黒87度
black 2 degrees　　　black 9 degrees　　　black 87 degrees

1) 黒色2度を含む実用版を1版、黒色9度を含む実用版を2版、黒色87度を含む実用版を3版に分類した。このうち、2版と3版は少ない。
1) The plates are classified by the screen angle of the black plate: 2° is plate 1, 9° is plate 2, and 87° is plate 3. Plates 2 & 3 are scarce.

C421 ● 定常変種 Constant Flaws

P1-A-pos.9　P1-B-pos.11　P1-C-pos.5　P1-C-pos.10

P1-D-pos.2　P1-D-pos.14　P2-A-pos.11　P3-B-pos.14

位置 position	特徴 feature	**	●
P1-A-pos.9	"郵"の下方に大きなリタッチ large retouch below "郵"	200	100
P1-B-pos.11	"東"の左上に白点 white dot to upper left of "東"		
P1-C-pos.5	"0"の下に黒い傷 black scratch below "0"		
P1-C-pos.10	車両の上方に白点2個、右マージンに青点 two white dots above train, blue dot in right margin		
P1-D-pos.2	"通"に黒点2個 two black dots beside "通"		
P1-D-pos.14	"開"の上方に大きい黒点 large black dot above "開"		
P2-A-pos.11	先端部に小さな青点 small blue dot on bullet train nose		
P3-B-pos.14	先端部に黒点 black dot on bullet train nose		

C422 国際文通週間　発行日：1964（昭和39）.10.4
International Letter Writing Week 1964　Date of Issue : 1964 (Showa 39) .10.4

C422 葛飾北斎画「富嶽三十六景 東海道程ヶ谷」
"Thirty-six Views of Mount Fuji : Hodogaya on the Tokaido" by Katsushika Hokusai

● 特印 Special Date Stamp

横浜 Yokohama

		**	●	FDC
C422 40 yen 多色 multicolored		160	100	550

切手データ Basic Information	
版式：グラビア輪転版	printing plate : rotary photogravure plate
目打：櫛型13	perforation : comb perf. 13
印面寸法：35.5×25mm	printing area size : 33.5×25mm
発行枚数：950万枚	number issued : 9.5 million
シート構成：10枚（横5×縦2）	pane format : 10 subjects (5x2)
製版・目打組合せ：タイプ1F	plate and perforation configuration : type 1F

C422 ● 定常変種 Constant Flaws

A-pos.1　B-pos.3　
　　　　　　　　　C-pos.5
　　　　　　　　　C-pos.6

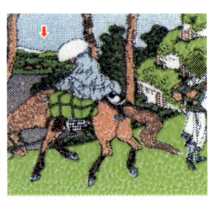
D-pos.5

E-pos.1

E-pos.10

F-pos.10

位置 position	特徴 feature	**	●
A-pos.1	落款の左に小さい紫点、松の右に青点 small purple dot to left of the signature, blue dot to right of pine tree	400	200
B-pos.3	"郵"の左に紫点 purple dot to left of "郵"		
C-pos.5	"0"の下に紫点3個 three purple dots below "0"		
C-pos.6	馬引きの頭上に大きな青点 large blue dot above head		
D-pos.5	笠の左上に小さい緑点 small green dot to upper left of bamboo hat		
E-pos.1	"便"の右に紫点2個 two purple dots to right of "便"		
E-pos.10	"国"の右に小さい紫点 small purple dot to right of "国"		
F-pos.10	縦長の青い破線 vertical blue dashed line		

記念・特殊切手（額面「円位」時期、ローマ字なし）

C423 富士山頂気象レーダー完成記念 発行日：1965（昭和40）.3.10
Meteorological Radar Station on Mt. Fuji Date of Issue：1965 (Showa 40) .3.10

C423 剣ヶ峰の富士山気象レーダー
Kengamine Meteorological Radar Station on Mt. Fuji

●特印
Special Date Stamp

富士宮 Fujinomiya

	**	●	FDC
C423 10 yen 多色 multicolored	50	30	200

C423 ●定常変種 Constant Flaws

P1-A-pos.2　P1-B-pos.5　P2-A-pos.11　P2-B-pos.11

C-pos.1　C-pos.14　D-pos.10　D-pos.13

切手データ Basic Information	
版式：グラビア輪転版	printing plate : rotary photogravure plate
目打：櫛型13½	perforation : comb perf. 13½
印面寸法：33×22.5mm	printing area size : 33×22.5mm
発行枚数：2,400万枚	number issued : 24 million
シート構成：20枚（横5×縦4）	pane format : 20 subjects (5x4)
製版・目打組合せ：タイプ2B-2	plate and perforation configuration : type 2B-2

位置 position	特徴 feature	**	●
P1-A-pos.2	"9"の右下に赤点 red dot to lower right of "9"	200	100
P1-B-pos.5	"6"の下に赤点 red dot below "6"		
P2-A-pos.11	"富"の左上に大きな黒点 large black dot to upper left of "富"		
P2-B-pos.11	"郵"の左に黒点 black dot to left of "郵"		
C-pos.1	レーダーの左に赤点 red dot to left of radar station		
C-pos.14	レーダーの左上に大きな白抜け large white spot to upper left of radar station		
D-pos.10	左マージン付近に青の縦線 vertical blue band near left margin		
D-pos.13	"6"の下方に赤点 red dot below "6"		

1) 赤、黒、青の各実用版は少なくとも2種あり、赤色の定常変種が多く出現するものを1版、黒色の定常変種が多く出現するものを2版に分類した。
1) The respective number of red, black and blue plates is at least two. The plate with the larger number of constant flaws in red is called plate 1, and the plate with larger number of constant flaws in black is called plate 2.

C424 逓信総合博物館竣工記念 世界切手展 発行日：1965（昭和40）.3.25
Completion of Communications Museum and World Stamp Exhibition Date of Issue：1965 (Showa 40) .3.25

C424 逓信総合博物館
Communications Museum, Tokyo

●特印
Special Date Stamp

東京 Tokyo

	**	●	FDC
C424 10 yen 暗い緑 green	50	30	300

切手データ Basic Information	
版式：グラビア輪転版	printing plate : rotary photogravure plate
目打：櫛型13½	perforation : comb perf. 13½
印面寸法：33×22,5mm	printing area size : 33×22.5mm
発行枚数：2,400万枚	number issued : 24 million
シート構成：20枚（横5×縦4）	pane format : 20 subjects (5x4)
製版・目打組合せ：タイプ2B-2	plate and perforation configuration : type 2B-2

C424
●定常変種 Constant Flaws

A-pos.6　A-pos.16　B-pos.3　B-pos.17

C-pos.8　C-pos.19　D-pos.9　D-pos.14

位置 position	特徴 feature	**	●
A-pos.6	"5"の右下に緑点 green dot to lower right of "5"	200	100
A-pos.16	博物館の上方に緑点 green dot above museum		
B-pos.3	"本"の左上に点、"本"の右下方に長い線 dot to upper left of "本", long line to lower right of "本"		
B-pos.17	"9"の一部欠け nick in "9"		
C-pos.8	博物館の上に緑点 green dot above museum		
C-pos.19	博物館の上に緑点 green dot above museum		
D-pos.9	博物館に右上に濃い点 dark dot to upper right of museum		
D-pos.14	"便"の下方に緑点 green dot below "便"		

87

記念・特殊切手（額面「円位」時期、ローマ字なし）

C425 切手趣味週間 / Philatelic Week 1965

発行日：1965（昭和40）.4.20　Date of Issue : 1965 (Showa 40) .4.20

C425 上村松園画「序の舞」
Japanese Dance "Jo-no-mai (The Prelude)" by Uemura Shoen

●特印 Special Date Stamp

東京 Tokyo

	**	●	FDC
C425 10 yen 多色 multicolored	60	50	400

切手データ Basic Information

版式：グラビア輪転版	printing plate : rotary photogravure plate
目打：櫛型13½	perforation : comb perf. 13½
印面寸法：33×48mm	printing area size : 33×48mm
発行枚数：3,800万枚	number issued : 38 million
シート構成：10枚（横5×縦2）	pane format : 10 subjects (5x2)
製版・目打組合せ：タイプ2C-2	plate and perforation configuration : type 2C-2

1) 目打は、二連型の櫛歯が中段縦1列の針を除いたものが使用された。
2) C425以降は各色のスクリーン角度が全色45度に統一されたことにより、各色の実用版を特定することは難しいため、定常変種により3面ずつを便宜的に1版、2版に分類した。

1) The perforator for this stamp was the double row comb perforator without the middle vertical row of pins.
2) Since the screen angle of all colors was unified to 45° for commemorative stamps after C425, it is difficult to identify the plate combination of each color. Therefore, for convenience, the plates are classified into plates 1 and 2 of three panes each according to the constant flaws.

C425 ●定常変種 Constant Flaws

P1-A-pos.8　P1-B-pos.8　P1-C-pos.2

P2-A-pos.8　P2-B-pos.3　P2-C-pos.8

位置 position	特徴 feature	**	●
P1-A-pos.8	裾に白抜け white spot on sleeve	200	100
P1-B-pos.8	"0" の下に赤点2個 two red dots below "0"		
P1-C-pos.2	袖の下に青点 blue dot below sleeve		
P2-A-pos.8	袖に大きな青点2個、裾の上の青点 two large blue dots on sleeve, blue dot above hem		
P2-B-pos.3	"65" の下に赤点 red dot below "65"		
P2-C-pos.8	着物に赤点 red dot on Kimono		

C426 国立こどもの国開園記念 / Opening of KODOMONOKUNI (Children's Land)

発行日：1965（昭和40）.5.5　Date of Issue : 1965 (Showa 40) .5.5

C426 遊ぶ子どもと動物、こどもの国のマーク Children Playing, Animals and KODOMONOKUNI Emblem

●特印 Special Date Stamp

横浜 Yokohama

	**	●	FDC
C426 10 yen 多色 multicolored	50	30	200

切手データ Basic Information

版式：グラビア輪転版	printing plate : rotary photogravure plate
目打：櫛型13½	perforation : comb perf. 13½
印面寸法：27×22.5mm	printing area size : 27×22.5mm
発行枚数：2,400万枚	number issued : 24 million
シート構成：20枚（横5×縦4）	pane format : 20 subjects (5x4)
製版・目打組合せ：タイプ2B-2	plate and perforation configuration : type 2B-2

1) 便宜的に1版と2版に分類したが、定常変種が確認できるのは赤色版と青色版がほとんどである。

1) The plates of this stamp are classified into plate 1 and plate 2 for convenience. Most of the constant flaws are found in the red and blue plates.

C426 ●定常変種 Constant Flaws

P1/P2-A-pos.18　P1-A-pos.9　P2-A-pos.2　P1-B-pos.13

P1/P2-C-pos.11　P1-C-pos.3　P2-C-pos.20　P1-D-pos.2　P2-D-pos.17

位置 position	特徴 feature	**	●
P1/P2-A-pos.18	牛の腹部に青点 blue dot on cow's abdomen	200	100
P1-A-pos.9	白鳥に青点 blue dot on swan		
P2-A-pos.2	"国" の右上方に濃い赤点 dark red dot to upper right of "国"		
P1-B-pos.13	"本" の左に大きい白点、上マージンに赤点 Large white dot to left of "本", red dot in top margin		
P1/P2-C-pos.11	草地の上に白点 white dot above grassland		
P1-C-pos.3	"こ" の上方に濃い赤点 dark red dot above "こ"		
P2-C-pos.20	女性の髪に青点 blue dot in woman's hair		
P1-D-pos.2	"立" の上方に大きい赤点 large red dot above "立"		
P2-D-pos.17	風車の上に濃い赤点 dark red dot above windmill		

記念・特殊切手（額面「円位」時期、ローマ字なし）

C427 国土緑化運動　　発行日：1965（昭和40）.5.9
National Afforestation Campaign　　Date of Issue : 1965 (Showa 40) .5.9

C427 樹木と陽光 Stylized Tree and Sun

●特印
Special Date Stamp

米子 Yonago

	**	●	FDC
C427 10 yen 多色 multicolored	50	30	200

切手データ Basic Information	
版式：グラビア輪転版	printing plate : rotary photogravure plate
目打：櫛型13½	perforation : comb perf. 13½
印面寸法：27×31mm	printing area size : 27×31mm
発行枚数：2,400万枚	number issued : 24 million
シート構成：20枚（横4×縦5）	pane format : 20 subjects (4×5)
製版・目打組合せ：タイプ1C-5	plate and perforation configuration : type 1C-5

C427 ●定常変種 Constant Flaws

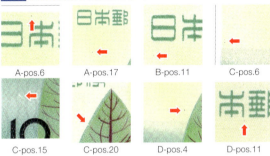

位置 position	特徴 feature	**	●
A-pos.6	"本"の上に黒点 black dot above "本"	200	100
A-pos.17	"日"の右下方に黒点 black dot to lower right of "日"		
B-pos.11	"日"の左下に緑点 green dot to lower left of "日"		
C-pos.6	左マージン付近に大小2個の黒点 two large and small black dots near left margin		
C-pos.15	"10"の上方に白抜け white dot above "10"		
C-pos.20	樹木の欠け chip in tree		
D-pos.4	樹木の左に黒点2個と赤点 two black dots and red dot to left of tree		
D-pos.11	"本"の右下に赤点 red dot to lower right of "本"		

C428 国際電気通信連合（ITU）100年記念　　発行日：1965（昭和40）.5.17
International Telecommunication Union (ITU) Centenary　　Date of Issue : 1965 (Showa 40) .5.17

C3428 電気通信100年の歩みと地球
100 years of Telecommunication and Globe

●特印
Special Date Stamp

東京 Tokyo

	**	●	FDC
C428 10 yen 紫味青・うす黄・黒 bright blue, yellow and black	50	30	200

切手データ Basic Information	
版式：グラビア輪転版	printing plate : rotary photogravure plate
目打：櫛型13½	perforation : comb perf. 13½
印面寸法：33×22.5mm	printing area size : 33×22.5mm
発行枚数：2,400万枚	number issued : 24 million
シート構成：20枚（横5×縦4）	pane format : 20 subjects (5×4)
製版・目打組合せ：タイプ2B-2	plate and perforation configuration : type 2B-2

C428 ●定常変種 Constant Flaws

A-pos.5　　A-pos.11　　B-pos.9　　B-pos.20

C-pos.12　　C-pos.19　　pos.8　　D-pos.19

位置 position	特徴 feature	**	●
A-pos.5	"日"の下に白点 white dot below "日"	200	100
A-pos.11	"便"の右に白い斜線 white diagonal line to right of "便"		
B-pos.9	アンテナの下に白点 white dot below antenna		
B-pos.20	"日"の上方に白点 white dot above "日"		
C-pos.12	地球の右に白点 white dot to right of globe		
C-pos.19	電線の上に白点 white dot above electrical wire		
D-pos.8	"電"の下に白点と白い傷 white dot and white scratch below "電"		
D-pos.19	"便"の右下方に白点 white dot to lower right of "便"		

記念・特殊切手（額面「円位」時期、ローマ字なし）

C429 国際協力年　発行日：1965（昭和40）.6.26
International Cooperation Year (ICY)　Date of Issue : 1965 (Showa 40) .6.26

	**	○	FDC
C429 40 yen 多色 multicolored	100	40	350

C429 協力年マークと平和のハト
ICY Emblem and Doves

●特印 Special Date Stamp

東京 Tokyo

切手データ Basic Information	
版式：グラビア輪転版	printing plate : rotary photogravure plate
目打：櫛型13½	perforation : comb perf. 13½
印面寸法：35.5×25mm	printing area size : 35.5×25mm
発行枚数：1,000万枚	number issued : 10 million
シート構成：10枚（横5×縦2）	pane format : 10 subjects (5x2)
製版・目打組合せ：タイプ1F	plate and perforation configuration : type 1F

C429 ●定常変種 Constant Flaws

A-pos.9

B-pos.2

C-pos.8

D-pos.7

D-pos.9

E-pos.9

F-pos.3

位置 position	特徴 feature	**	○
A-pos.9	"国"の下に白点、"6"の下方に白点、ハトの周辺に複数の白点 white dot below "国", white dot below "6", multiple white dots around dove	200	100
B-pos.2	ハトに黒点、"CO"の上に白点 black dot on dove, white dot above "CO"		
C-pos.8	ハトに赤点 red dot on dove		
D-pos.7	"Y"の左に白点 white dot to left of "Y"		
D-pos.9	"国"の下方に白点、"本"の左下方に白点 white dot below "国", white dot to lower left of "本"		
E-pos.9	ハトに金色点2個 two gold dots on dove		
F-pos.3	ハトに黒点 black dot on dove		

C430 第25回海の記念日　発行日：1965（昭和40）.7.20
25th Marine Day　Date of Issue : 1965 (Showa 40) .7.20

	**	○	FDC
C430 10 yen 多色 multicolored	50	30	200

C430 明治丸とウミネコ
Meiji Maru and Black-tailed Gulls

●特印 Special Date Stamp

横浜 Yokohama

切手データ Basic Information	
版式：グラビア輪転版	printing plate : rotary photogravure plate
目打：櫛型13½	perforation : comb perf. 13½
印面寸法：33×22.5mm	printing area size : 33×22.5mm
発行枚数：2,500万枚	number issued : 25 million
シート構成：20枚（横5×縦4）	pane format : 20 subjects (5x4)
製版・目打組合せ：タイプ2B-2	plate and perforation configuration : type 2B-2

C430 ●定常変種 Constant Flaws

A-pos.9

A-pos.20

位置 position	特徴 feature	**	○
A-pos.9	"1"の上の海面欠け nick in sea surface above "1"	200	100
A-pos.20	左上のウミネコの下に黒点、右マージンに緑点 black dot below black-tailed gull, green dot in right margin		
B-pos.1	左上のウミネコの翼右にリタッチ retouch to right of left black-tailed gull's wing		
C-pos.16	左上のウミネコの上に濃い紫点 dark purple dot above left black-tailed gull		
C-pos.19	船先の左方に緑のシミ green stain to left of ship's prow		
D-pos.18	左上のウミネコの翼上に小さな緑点 small green dot above left black-tailed gull's wing		
D-pos.20	左上のウミネコの右上方に濃い緑点 dark green dot to upper right of left black-tailed gull		

B-pos.1

拡大図 enlarged stamp

修正版 revised version

C-pos.16

C-pos.19

D-pos.18

D-pos.20

記念・特殊切手（額面「円位」時期、ローマ字なし）

C431　愛の血液助け合い運動
Blood Services
発行日：1965（昭和40）.9.1
Date of Issue：1965 (Showa 40).9.1

C431 血液中の子どもと採血車
Child in blood drop and Bloodmobile

●特印
Special Date Stamp

東京 Tokyo

	**	●	FDC
C431 10 yen 多色 multicolored	50	30	200

切手データ Basic Information

版式：グラビア輪転版	printing plate : rotary photogravure plate
目打：櫛型13½	perforation : comb perf. 13½
印面寸法：33×22.5mm	printing area size : 33×22.5mm
発行枚数：2,500万枚	number issued : 24 million
シート構成：20枚（横5×縦4）	pane format : 20 subjects (5x4)
製版・目打組合せ：タイプ2B-2	plate and perforation configuration : type 2B-2

C431 ●定常変種 Constant Flaws

A-pos.1

A-pos.15

B-pos.1

B-pos.9

C-pos.2
C-pos.15

D-pos.6
D-pos.10

位置 position	特徴 feature	**	●
A-pos.1	"合"の右下に白抜け white spot to lower right of "合"	200	100
A-pos.15	採血車の右下に白抜け white spot to lower right of bloodmobile		
B-pos.1	左マージン近くに赤点 red dot near left margin		
B-pos.9	"日"の左方に大きな緑点 large green dot to left of "日"		
C-pos.2	"本"に黒点 black dot in "本"		
C-pos.15	採血車の下方に赤点と白点 red dot and white dot below bloodmobile		
D-pos.6	"動"の下方に白抜け white spot below "動"		
D-pos.10	"1"の左に濃い緑のシミ dark green stain to left of "1"		

C432　第9回国際原子力機関総会記念
9th Conference of International Atomic Energy Agency (IAEA)
発行日：1965（昭和40）.9.21
Date of Issue：1965 (Showa 40).9.21

C432 東海村の原子力発電所
Tokai Village Nuclear Power Plant

●特印
Special Date Stamp

東京 Tokyo

	**	●	FDC
C432 10 yen 多色 multicolored	50	30	200

切手データ Basic Information

版式：グラビア輪転版	printing plate : rotary photogravure plate
目打：櫛型13½	perforation : comb perf. 13½
印面寸法：27×22.5mm	printing area size : 27×22.5mm
発行枚数：2,500万枚	number issued : 25 million
シート構成：20枚（横5×縦4）	pane format : 20 subjects (5x4)
製版・目打組合せ：タイプ2B-2	plate and perforation configuration : type 2B-2

1) 便宜的に1版と2版に分類したが、定常変種が確認できるのは全て青色版である。全般にベタ地の各所に白点が見られるが、ほとんどが偶発的なもので注意を要する。

1) The plates of this stamp are classified into plate 1 and plate 2 for convenience. All the constant flaws are found in the blue plate. Generally, white spots can be seen in various places on the solid blue ground, but most are random occurrences and warrant caution.

C432 ●定常変種 Constant Flaws

P1-A-pos.13

P1-B-pos.19

P1-C-pos.3

P2-A-pos.19
P2-B-pos.19

P1-D-pos.19

位置 position	特徴 feature	**	●
P1-A-pos.13	"日"の左下に白点 white dot under left of "日"	200	100
P1-B-pos.19	"郵"の下方に白点、"製"の下に青点 white dot below "郵", blue dot under "製"		
P1-C-pos.3	リタッチ状の濃い青点 retouched dark blue dot		
P1-D-pos.19	"郵"の右下方に白点、"省"の右上に青点 White dot below right of "郵", blue dot above right of "省"		
P2-A-pos.19	銘版の上部全体に青の破線 dashed blue line across top of imprint		
P2-B-pos.19	銘版の上部全体に青の破線 dashed blue line across top of imprint		

記念・特殊切手（額面「円位」時期、ローマ字なし）

C433 第10回国勢調査記念　発行日：1965（昭和40）.9.25
10th Census　Date of Issue : 1965 (Showa 40) .9.25

C433 日の丸と人口を示すこけし
Japanese Flag and Kokeshi dolls symbolizing population

●特印 Special Date Stamp

東京 Tokyo

		**	●	FDC
C433 10 yen 多色 multicolored		50	30	200

切手データ Basic Information	
版式：グラビア輪転版	printing plate : rotary photogravure plate
目打：櫛型13½	perforation : comb perf. 13½
印面寸法：33×22.5mm	printing area size : 33×22.5mm
発行枚数：2,500万枚	number issued : 25 million
シート構成：20枚（横5×縦4）	pane format : 20 subjects (5x4)
製版・目打組合せ：タイプ2B-2	plate and perforation configuration : type 2B-2

C433 ●定常変種 Constant Flaws

位置 position	特徴 feature	**	●
A-pos.12	"国"の上方に白点 white dot above "国"	200	100
A-pos.18	"便"の右下に白点 white dot to lower right of "便"		
B-pos.8	"勢"の上方に白抜け white spot above "勢"		
B-pos.16	"1"の右に白点 white dot to right of "1"		
C-pos.4	こけしに濃い紫点 dark purple dot on Kokeshi doll		
C-pos.9	"0"の下に黒点 black dot below "0"		
D-pos.13	下マージンに赤点 red dot in bottom margin		

A-pos.12　A-pos.18　B-pos.8　B-pos.16
C-pos.4　C-pos.9　D-pos.13

C434 国際文通週間　発行日：1965（昭和40）.10.6
International Letter Writing Week 1965　Date of Issue : 1965 (Showa 40) .10.6

C434 葛飾北斎画「富嶽三十六景 甲州三坂水面」
"Thirty-six Views of Mount Fuji :
Fuji Reflected in Lake at Misaka in Kai Province"
by Katsushika Hokusai

●特印 Special Date Stamp
富士吉田 Fujiyoshida

		**	●	FDC
C434 40 yen 多色 multicolored		150	100	450

切手データ Basic Information	
版式：グラビア輪転版	printing plate : rotary photogravure plate
目打：全型13	perforation : harrow perf. 13
印面寸法：35.5×25mm	printing area size : 33.5×25mm
発行枚数：1,200万枚	number issued : 12 million
シート構成：10枚（横5×縦2）	pane format : 10 subjects (5x2)
製版・目打組合せ：タイプ2D	plate and perforation configuration : type 2D

C434 ●定常変種 Constant Flaws

P1/P2-A-pos.5　P1-A-pos.8　P2-A-pos/2　P1-B-pos.3

P1-C-pos.5　P2-C-pos.7

P1-D-pos.3　P1-E-pos.10

P2-B-pos.2

P1-F-pos.4

1) 多色刷グラビア輪転版により6面縦並びで印刷された後に全型目打で連続穿孔された最初の切手で、1版、2版で9面分が確認されている。
1) This was the first stamp printed on the multi-color rotary photogravure press with a six-pane vertical configuration plate, followed by the continuous harrow perforation. Nine different panes have been identified for plates 1 and 2.

記念・特殊切手（額面「円位」時期、ローマ字なし）

位置 position	特徴 feature	**	●
P1/P2-A-pos.5	"LETTER"の上の印面にリタッチ retouch above "LETTER"	300	200
P1-A-pos.8	上マージンに青点 blue dot in top margin		
P2-A-pos.2	湖面の富士山に青点 blue dot on Mt. Fuji's reflection		
P1-B-pos.3	砂浜に青点2ヵ所 two blue dots on lake shore		

位置 position	特徴 feature
P2-B-pos.2	"4"の左に青点、"国"の左に青点 blue dot to left of "4", blue dot to left of "国"
P1-C-pos.5	"0"の中に青点 blue dot in "0"
P2-C-pos.7	"G"の右上方に青点 blue dot to upper right of "G"
P1-D-pos.3	小舟の下方に青点 blue dot below boat
P1-E-pos.10	湖面の富士山に大きい青点 large blue dot on Mt. Fuji's reflection
P1-F-pos.4	富士山の上に青いシミ blue stain above Mt. Fuji

C435 国民参政75周年記念　発行日：1965（昭和40）.10.15
75th Anniversary of General Suffrage　Date of Issue：1965 (Showa 40).10.15

C435 国会の議席と議員の記章
National Diet Seating Diagram and Diet Members' Badges

●特印
Special Date Stamp

東京 Tokyo

	**	●	FDC
C435 10 yen 多色 multicolored	50	30	200

切手データ Basic Information	
版式：グラビア輪転版	printing plate : rotary photogravure plate
目打：櫛型13½	perforation : comb perf. 13½
印面寸法：27×22.5mm	printing area size : 27×22.5mm
発行枚数：2,500万枚	number issued : 25 million
シート構成：20枚（横5×縦4）	pane format : 20 subjects (5x4)
製版・目打組合せ：タイプ2B-2	plate and perforation configuration : type 2B-2

C435 ● 定常変種 Constant Flaws

A-pos.6　　A-pos.7

 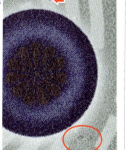
B-pos.20　　B-pos.15

C-pos.10　C-pos.19　D-pos.2　D-pos.14

位置 position	特徴 feature	**	●
A-pos.6	青の記章に白点と濃い青点 white dot and dark blue dot on blue badge	200	100
A-pos.7	青の記章に白点 white dot on blue badge		
B-pos.15	青の記章の上に小さい緑点と記章の下に濃い点群 small green dot above blue emblem, dark dots below badge		
B-pos.20	"0"の上に赤紫点 red-purple dot above "0"		
C-pos.10	第1コーナーのマージンに青点 blue dot in margin of NW corner		
C-pos.19	"1"の左上方に赤紫点 red-purple dot to upper left of "1"		
D-pos.2	"40"の右上方に赤紫点 red-purple dot to upper right of "40"		
D-pos.14	郵の下方に濃い緑点 dark green dot below "郵"		

C436-437 第20回国体記念　発行日：1965（昭和40）.10.24
20th National Athletic Meet　Date of Issue：1965 (Showa 40).10.24

C436 あん馬　　C437 競歩
Pommel Horse　Race Walking

●特印
Special Date Stamp

岐阜 Gifu

	**	●	FDC
C436 5 yen 赤茶 red-brown	40	30	
C437 5 yen オリーブ yellow-green	40	30	
C436-437 (2種連刷 pair)	90	100	180

切手データ Basic Information	
版式：ザンメル凹版	printing plate : sammeldruck intaglio printing
目打：櫛型13½	perforation : comb perf. 13½
印面寸法：22.5×27mm	printing area size : 22.5×27mm
発行枚数：各2,500万枚	number issued : 25 million each
シート構成：20枚（横4×縦5）	pane format : 20 subjects (4x5)
製版・目打組合せ：タイプ1C-5	plate and perforation configuration : type 1C-5

記念・特殊切手（額面「円位」時期、ローマ字なし）

C438 国際耳鼻咽喉・小児科学会議記念
International Otorhinolaryngology and Pediatric Congress

発行日：1965（昭和40）.10.30
Date of Issue : 1965 (Showa 40) .10.30

C438 耳鼻咽喉を表す横顔と小児科を表す幼児
Profile Representing Otorhinolaryngology and Infant Representing Pediatrics

●特印 Special Date Stamp

東京 Tokyo

	**	●	FDC
C438 10 yen 多色 multicolored	80	40	200

切手データ Basic Information	
版式：グラビア輪転版	printing plate : rotary photogravure plate
目打：全型13½	perforation : harrow perf. 13½
印面寸法：35.5×25mm	printing area size : 35.5×25mm
発行枚数：1,200万枚	number issued : 12 million
シート構成：10枚（横5×縦2）	pane format : 10 subjects (5x2)
製版・目打組合せ：タイプ2D	plate and perforation configuration : type 2D

C438 ● 定常変種 Constant Flaws

 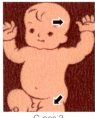

A-pos.4　A-pos.5　B-pos.3　C-pos.3

D-pos.2　E-pos.4　E-pos.5　F-pos.1

position	特徴 feature	**	●
A-pos.4	耳の左に白ぼやけ white blur to left of ear	200	100
A-pos.5	下マージンに赤点 red dot in bottom margin		
B-pos.3	幼児の胸に赤点 red dot on infant's chest		
C-pos.3	頭に大きい白点、太腿に赤点 large white dot on head, red dot on thigh		
D-pos.2	"P"の中に赤点 red dot in "P"		
E-pos.4	手首に濃い赤点 dark red dot on wrist		
E-pos.5	掌に大きな白点 large white spot on palm		
F-pos.1	右足に赤点 red dot on right foot		

C439 南極地域観測再開記念
Resumption of Antarctic Observation

発行日：1965（昭和40）.11.20
Date of Issue : 1965 (Showa 40) .11.20

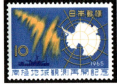

C439 南極地図、オーロラ、観測船「ふじ」
Antarctic map, Aurora and Observation Ship "Fuji"

●特印 Special Date Stamp

東京 Tokyo

	**	●	FDC
C439 10 yen 青・暗い青・赤味黄 blue, dark blue and yellow	50	30	200

切手データ Basic Information	
版式：グラビア輪転版	printing plate : rotary photogravure plate
目打：櫛型13½	perforation : comb perf. 13½
印面寸法：33×22.5mm	printing area size : 33×22.5mm
発行枚数：2,550万枚	number issued : 25.5 million
シート構成：20枚（横5×縦4）	pane format : 20 subjects (5x4)
製版・目打組合せ：タイプ2B-2	plate and perforation configuration : type 2B-2

C439 ● 定常変種 Constant Flaws

A-pos.3　B-pos.6　B-pos.8

  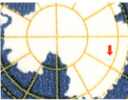

C-pos.18　C-pos.19　D-pos.3　D-pos.11/16

position	特徴 feature	**	●
A-pos.3	昭和基地の左に大きな青点 large blue blot to left of Showa Base	200	100
B-pos.6	南極地図の左上部に青点2個 two blue dots in on upper part of Antarctica map		
B-pos.8	南極地図の右下部に青点 blue dot on lower right part of Antarctica map		
C-pos.18	"0"の右に白点 white dot to right of "0"		
C-pos.19	"南"の下に青点 blue dot below "南"		
D-pos.3	第2コーナーに青点 blue dot in NE corner		
D-pos.11/16	"極"の下に黄点、上マージンに青点 yellow dot below "極", blue dot in top margin		

記念・特殊切手（額面「円位」時期、ローマ字なし／ローマ字入り）

C440 電話創業75年記念　発行日：1965（昭和40）.12.16
75th Anniversary of Telephone Service　Date of Issue: 1965 (Showa 40).12.16

C440 ダイヤルと創業当時の電話交換機
Telephone Dial and 1890s Switchboard

●特印 Special Date Stamp

東京 Tokyo

	**	●	FDC
C440 10 yen 多色 multicolored	50	30	200

切手データ Basic Information	
版式：グラビア輪転版	printing plate: rotary photogravure plate
目打：櫛型13½	perforation: comb perf. 13½
印面寸法：22.5×27mm	printing area size: 22.5×27mm
発行枚数：2,500万枚	number issued: 25 million
シート構成：20枚（横4×縦5）	pane format: 20 subjects (4×5)
製版・目打組合せ：タイプ2B-1	plate and perforation configuration: type 2B-1

C440 ● 定常変種 Constant Flaws

A-pos.7　A-pos.18　B-pos.6　B-pos.20

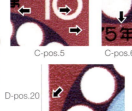

C-pos.5　C-pos.6

D-pos.20

D-pos.8

位置 position	特徴 feature	**	●
A-pos.7	ダイヤルに紫点 purple dot in telephone dial hole	200	100
A-pos.18	"本"の下に白点2個 two white dots below "本"		
B-pos.6	第4コーナーの上部欠け nick in background at SW corner		
B-pos.20	第1コーナーのマージンに黒点 black dot in margin of NW corner		
C-pos.5	"便"の下に白点、"0"の中と右下方に白点 white dot below "便", white dots inside and below right of "0"		
C-pos.6	"年"の上に白点 white dot above "年"		
D-pos.8	ダイヤルの左に大きい白点、"1"の左に白点、"0"の右下方に白点と白もや large white dot to left of dial, white dot to left of "1", white dot and white haze to lower right of "0"		
D-pos.20	ダイヤルの左に白抜け nick in background to left of telephone dial		

(3) 額面「円位」時期、ローマ字入り "Yen" Value Era, with NIPPON 1966

C441-445 魚介シリーズ　発行年：1966（昭和41）
Fish and Shellfish Series　Year of Issue: 1966 (Showa41)

C441 イセエビ
Japanese Spiny Lobster

C442 コイ Carp

C443 マダイ Sea Bream

●風景印 Pictrial Postmarks

C441 牛深　C442 野沢　C443 鞆
Ushibuka　Nozawa　Tomo

C444 カツオ Skipjack Tuna

C445 アユ Sweetfish

C443 小湊　C444 焼津　C444 土佐清水　C445 郡上八幡
Kominato　Yaizu　Tosashimizu　Gujo Hachiman

記念・特殊切手（額面「円位」時期、ローマ字入り）

			**	●	FDC
C441	10 yen 多色 multicolored (1966.1.31)	……50	50	350
C442	10 yen 多色 multicolored (1966.2.28)	……50	30	350
C443	10 yen 多色 multicolored (1966.3.25)	……50	30	350
C444	10 yen 多色 multicolored (1966.5.16)	……50	30	350
C445	10 yen 多色 multicolored (1966.6.1)	……50	30	300

切手データ Basic Information

版式：グラビア輪転版	printing plate : rotary photogravure plate
目打：櫛型13½	perforation : comb perf. 13½
印面寸法：各35.5×25mm	printing area size : 35.5×25mm each
発行枚数＝C441-443＝各2,800万枚、C444-445＝各3,000万枚	number issued : C441-443＝28 million each, C444-445＝30 million each
シート構成：各20枚（横5×縦4）	pane format : 20 subjects (5x4) each
製版・目打組合せ：タイプ1D-3	plate and perforation configuration : type 1D-3

C441 ●定常変種 Constant Flaws

A-pos.18　　B-pos.16　　C-pos.10

D-pos.6　　D-pos.16　　D-pos.18

位置 position	特徴 feature	**	●
A-pos.18	脚の下に白点 white dot below legs	200	100
B-pos.16	第3コーナーのマージンに青点 blue dot in margin of SE corner		
C-pos.10	"1"の左に赤点、第3コーナーのマージンに青点 red dot to left of "1", blue dot in margin of SE corner		
D-pos.6	ひげの上に白点 white dot above antennae		
D-pos.16	"日"の左上方に白抜き点 white spot to upper left "日"		
D-pos.18	"日"の上方に白抜き点 white spot above "日"		

C442 ●定常変種 Constant Flaws

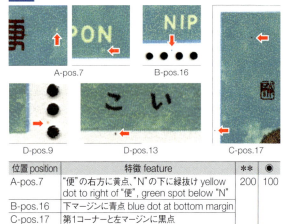

A-pos.7　　B-pos.16

D-pos.9　　D-pos.13　　C-pos.17

位置 position	特徴 feature	**	●
A-pos.7	"便"の右方に黄点、"N"の下に緑抜け yellow dot to right of "便", green spot below "N"	200	100
B-pos.16	下マージンに青点 blue dot at bottom margin		
C-pos.17	第1コーナーと左マージンに黒点 black dots in NW corner and left margin		

D-pos.9　　第3コーナーのマージンに赤点　red dot in margin of SE corner
D-pos.13　　"こ"の下方のマージンに黒点　black dot in margin below "こ"

C443 ●定常変種 Constant Flaws

A-pos.2　　B-pos.4/9

B-pos.20

C-pos.16　　C-pos.19　　D-pos.9

C-pos.14　　D-pos.16

位置 position	特徴 feature	**	●
A-pos.2	腹部に黒点2ヵ所、"い"の下方のマージンに青点 two black dots on abdomen, blue dot in margin bottom below "い"	200	100
B-pos.4/9	マージン中間に青点 blue dot in margin between positions 4 and 9		
B-pos.20	右マージンに青点 blue dot in right margin		
C-pos.16	左マージンに青点 blue dot in left margin		
C-pos.19	上マージンに青点 blue dot in top margin		
D-pos.9	右マージンに青点 blue dot in right margin		
D-pos.14	下マージンにオレンジ点 orange dot in bottom margin		
D-pos.16	左マージンに青点 blue dot in left margin		

C444 ●定常変種 Constant Flaws

A-pos.16

B-pos.14

右上につづく Continued to up right ↗

記念・特殊切手（額面「円位」時期、ローマ字入り）

C445 ●定常変種 Constant Flaws

C-pos.5

C-pos.10

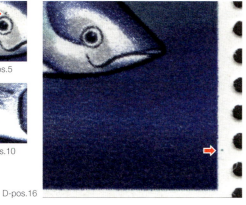
D-pos.16

A-pos.3

B-pos.18/19

C-pos.3

C-pos.7

C-pos.18

D-pos.1

D-pos.2

D-pos.11

位置 position	特徴 feature	**	●
A-pos.3	"0"の中に赤点 red dot on "0"	200	100
B-pos.18/19	マージンに赤点2ヵ所 two red dots in margin between positions 18 and 19		
C-pos.3	下マージンに小さな赤点 small red dot in bottom margin		
C-pos.7	"N"の左下方に大きな黒点 large black dot to lower left of "N"		
C-pos.18	左マージンに緑点 green dot in left margin		
D-pos.1	"P"の上のマージンに黒点 black dot in upper margin above "P"		
D-pos.2	腹びれの上にオレンジ点 orange dot above pelvic fin		
D-pos.11	右マージンに黒点 black dot in right margin		

位置 position	特徴 feature	**	●
A-pos.16	口の下方に青抜け点 blue spot below mouth	200	100
B-pos.14	エラから腹にかけて青点3ヵ所 three blue dots from gill to belly		
C-pos.5	目の上に濃い青点 dark blue dot above eye		
C-pos.10	腹に青点2個 two blue dots on belly		
D-pos.16	第3コーナーのマージンに緑点 green dot in margin at SE corner		

C453 名園シリーズ 偕楽園　発行日：1966（昭和41）.2.25
Famous Garden Series, Kairakuen Park　Date of Issue : 1966 (Showa 41) .2.25

C453 好文亭と白梅
Kobuntei Pavilion and White Plums

●風景印
Pictrial Postmarks

水戸 Mito

		**	●	FDC
C453 10 yen 赤金・こい緑・茶黒 gold, black and green		50	30	450

切手データ Basic Information	
版式：グラビア輪転版	printing plate : rotary photogravure plate
目打：櫛型13½	perforation : comb perf. 13½
印刷寸法：22.5×33mm	printing area size：22.5×33mm
発行枚数：2,400万枚	number issued：24 million
シート構成：20枚（横4×縦5）	pane format : 20 subjects (4×5)
製版・目打組合せ：タイプ2B-1	plate and perforation configuration：type 2B-1

C453 ●定常変種 Constant Flaws

A-pos.4

C-pos.3

B-pos.20

D-pos.4

D-pos.8

D-pos.11

位置 position	特徴 feature	**	●
A-pos.4	"水"の左下方に金の突起、"ON"の下に大きな白点 gold protrusion to lower left of "水", large white dot below "ON"	200	100
B-pos.20	"園"の右下方に黒点 black dot to lower right of "園"		
C-pos.3	左マージンに緑点 green dot in left margin		
D-pos.4	白梅に緑点 green dot on plum blossom		
D-pos.8	"0"の中に緑点 green dot inside "0"		
D-pos.11	白梅に緑点 green dot on plum blossom		

97

記念・特殊切手（額面「円位」時期、ローマ字入り）

C456 国際工業所有権保護協会東京総会記念
AIPPI (International Association for the Protection of Intellectual Property) Tokyo Congress

発行日：1966（昭和41）.4.11
Date of Issue : 1966 (Showa 41) .4.11

C456 特許、実用新案、意匠、商標のシンボルに協会マークsymbols for patents, utility models, designs and trademarks and emblem of AIPPI

●特印 Special Date Stamp

東京 Tokyo

	**	●	FDC
C456 10 yen 多色 multicolored	100	40	280

切手データ Basic Information	
版式：グラビア輪転版	printing plate : rotary photogravure plate
目打：全型13½	perforation : harrow perf. 13½
印面寸法：36×25mm	printing area size : 36×25mm
発行枚数：1,000万枚	number issued : 10 million
シート構成：10枚（横5×縦2）	pane format : 10 subjects (5x2)
製版・目打組合せ：タイプ2D	plate and perforation configuration : type 2D

C456 ●定常変種 Constant Flaws

 A-pos.8

 A-pos.10

 D-pos.8

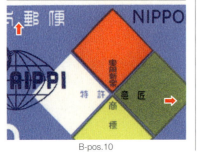 B-pos.10

C-pos.10

位置 position	特徴 feature	**	●
A-pos.8	"総会"の上方に白点 white dot above "総会"	200	100
A-pos.10	"便"の左上が白抜け white spot to upper left of "便"		
B-pos.10	"本"と"郵"の間に白点2カ所、"匠"の右方に白抜 two white dots between "本" and "郵", white spot to right of "匠"		
C-pos.10	"P"と"P"の間に白点 white dot between "P" and "P"		
D-pos.8	"4"と"際"の間に白点2ヵ所 two white dots between "4" and "際"		

C457 切手趣味週間
Philatelic Week 1965

発行日：1966（昭和41）.4.20
Date of Issue : 1966 (Showa 41) .4.20

C457 藤島武二画「蝶」 "Butterflies" by Fujishima Takeji

●特印 Special Date Stamp

東京 Tokyo

	**	●	FDC
C457 10 yen 多色 gold and multicolored	50	40	350

切手データ Basic Information	
版式：グラビア・局式凹版	printing plate : rotary photogravure plate with intaglio
目打：櫛型13½	perforation : comb perf. 13½
印面寸法：48×33mm	printing area size : 48×33mm
発行枚数：3,500万枚	number issued : 35 million
シート構成：10枚（横2×縦5）	pane format : 10 subjects (2x5)
製版・目打組合せ：タイプ2C-1	plate and perforation configuration : type 2C-1

1）3面縦並びの連続目打で穿孔されたタイプ2C-1で、「局式凹版」のユニットが連結されて髪の毛の部分が印刷された。
2）各色の実用版を特定することが難しいため、定常変種により3面ずつを便宜的に1版、2版に分類した。

1) The plate and perforation configuration of this stamp is type 2C-1: the plate is 3-pane vertical configuration and the perforation is continuous comb type. The woman's hair was printed with the printing bureau developed intaglio unit added to the photogravure press.
2) Since it is difficult to determine the plate combination of each color, the plates are classified into plates 1 and 2 of three panes each, according to the constant flaws, for convenience.

C457 ●定常変種 Constant Flaws

 P1-A-pos.5

 P1-B-pos.8

 P1-C-pos.2

 P2-A-pos.9

 P2-B-pos.1

 P2-C-pos.7

位置 position	特徴 feature	**	●
P1-A-pos.5	"郵"の左下に茶点 brown dot to lower left of "郵"	200	100
P1-B-pos.8	"1"の右上方に茶点 brown dot to upper right of "1"		
P1-C-pos.2	目の下に茶点 brown dot below eye		
P2-A-pos.9	"6"の左右に茶点 brown dots to left and right of "6"		
P2-B-pos.1	頬に茶点 brown dot on cheek		
P2-C-pos.7	"6"の右に茶点 brown dot to right of "6"		

原画作者 Designers

●郵政省技芸官 Designers of Ministry of Posts and Telecommunications

原画作者 Designers	カタログ番号 catalogue No.
大塚均 Otsuka Hitoshi	C284, C289, C291, C296, C316, C322, C326, C331, C337, C338, C348, C354, C357, C361, C374, C375, C376, C377, C390, C397, C402, C408, C410, C421, C423, C435, C436
加曾利鼎造 Kasori Teizo	C229, C234
木村勝 Kimura Masaru	C232, C233, C235, C248, C249, C252, C254, C256, C264, C266, C269, C270, C274, C276, C288, C290, C292, C298, C306, C307, C309, C321, C325, C342, C347, C352, C353, C362, C364, C383, C385, C386, C394, C400, C401, C403, C404, C405, C406, C411, C412, C426, C432
長谷部日出男 Hasebe Hideo	C255, C259, C277, C305, C333, C344, C349, C351, C355, C359, C363, C379, C384, C391, C393, C420, C425, C427, C430, C457
日置勝駿 Hioki Masatoshi	C300, C398, C440
東角井良臣 Higashitsunoi Yoshiomi	C328, C409, C428, C328
久野實 Hisano Minoru	C226, C227, C230, C236, C239, C243, C246, C260, C261, C263, C272, C273, C278, C281, C282, C286, C287, C293, C298, C302, C303, C315, C317, C324, C345, C358, C360, C366, C373, C380, C381, C389, C413, C429, C431, C424
山野内孝夫 Yamanouchi Takao	C301, C311, C312, C313, C437, C453
吉田豊 Yoshida Yutaka	C247, C257, C271, C308
渡邊三郎 Watanabe Saburo	C225, C231, C240, C241, C242, C244, C245, C250, C251, C258, C262, C265, C267, C268, C275, C279, C280, C283, C285, C295, C299, C304, C310, C314, C318, C320, C323, C339, C340, C341, C343, C346, C350, C356, C365, C378, C382, C387, C388, C395, C396, C399, C407, C414, C415, C416, C417, C418, C419, C422, C433, C434, C438, C456

●画家、彫刻家等 Painters, Sculptors, etc.

画家、彫刻家等 Painters, Sculptors, etc.	カタログ番号 catalogue No.
江守若菜 Emori Wakana	C392
加藤栄三 Kato Eizo	C441
堅山南風 Katayama Nanpu	C442
河野孝治 Kono Koji	C319
島田武夫 Shimada Tetsuo	C297, C330, C336
杉山寧 Shugiyama Yashushi	C445
丹後昭 Tango Akira	C329
綱島廉 Tsunajima Ren	C253
中尾竜作 Nakao Ryusaku	C237
中島桂 Nakajima Katsura	C334
野沢保人 Nozawa Yashuto	C332
乘松巖 Norimatsu Iwao	C439
橋本明治 Hashimoto Meiji	C444
前川治朗 Maekawa Jiro	C238
前田青邨 Maeda Seison	C443
前野京平 Maeno Kyohei	C327, C335
宮本三郎 Miyamoto Saburo	C228

凹版切手の原版彫刻者 Engraver of Intaglio Printing Stamps

原版彫刻者 Engraver of Intaglio Printing Stamps	カタログ番号 catalogue No.
岡村昭二 Okamura Shouji	C313, C344, C358, C402, C437
押切勝造 Oshikiri Katsuzou	C245, C250, C259, C267, C274, C282, C290, C292, C308, C312, C323, C345, C348, C350, C354, C357
笠野常雄 Kasano Tsuneo	C234, C247, C254, C266, C268, C284, C311, C321, C347, C352, C353, C356, C361, C364
河原義郎 Kawahara Yoshiro	C237, C251 C257, C258, C326, C363
栗原七三 Kurihara Shichizou	C256, C263, C270, C297, C300, C307, C349, C360, C365
中田昭吉 Nakata Shoukichi	C315, C322, C384, C401, C409, C436
長谷川秀男 Hasegawa Hideo	C232, C233, C362, C410
松浦匡夫 Matsuura Tadao	C229
水谷千吉 Mizutani Senkichi	C359, C366, C381, C383
渡部文雄 Watabe Fumio	C235, C244, C246, C248, C264, C283, C309, C341, C351, C355

使用例評価 Valuations by Cover

※使用例評価は、適正（適応）使用を基準としている。　※The value of covers are based on proper (adapted) usage.

C225 - C438 国民体育大会、切手趣味週間、国際文通週間、東京1964オリンピック競技大会（寄附金付）を除く Excluding National Athletic Meet, Philatelic Week, International Letter Writing Week and Tokyo 1964 Olympic Games（Semi-Postal）

●5 yen　C225, C227, C232, C237, C247, C255, C275, C289, C325　A：C275, C289, C318, C325, C414

地域 Area		郵便種別 Type of Mail	C225	C227	C237, C247	C255	A
国内郵便 domestic mail	1枚貼私製はがき	postcard bearing 5 yen stamp	2,500	2,500	1,000	800	500
	2枚貼書状	letter bearing two 5 yen stamps	2,500	2,500	1,500	1,000	800
	各種国内便	various domestic mail	2,000	2,000	1,000	800	500
外国郵便 international mail	1枚貼外信印刷物	printed matter bearing two 5 yen stamps		4,000	3,000	2,000	1,500
	5枚貼特別地帯宛航空便書状	air mail letter to special zone bearing five 5 yen stamps				3,000	3,000

記念・特殊切手（使用例評価）

●**10 yen**　A：C226, C228, C236, C238, C241-C244, C248　B：C249 - C271（excluding C255）
　　　C：C272 - C304（excluding C275, C277, C278, C289-C292）, C311, C319, C326, C415　D：C327, C328, C339　E：C341 - C373　F: C374 - C440

地域 Area		郵便種別 Type of Mail	A	B	C	D	E	F
国内郵便 domestic mail	1枚貼書状	letter bearing 10 yen stamp	2,000	1,000	500	1,000	500	400
	2枚貼2倍重量書状	double weight letter bearing two 10 yen stamps		1,500	800	1,500	800	600
	1枚貼印刷物	printed matter bearing 10 yen stamp	3,000		800	1,000	3,000	500
	各種国内便	various domestic mail	2,000	1,000	500	1,000	500	400
外国郵便 international mail	1枚貼外信印刷物	printed matter bearing 10 yen stamp	4,000	2,500	1,500	3,000	3,000	
	5枚貼特別地帯宛航空便2倍重量書状	air mail double weight letter to the special zone bearing five 10 yen stamps			3,000	4,000	3,000	2,500
	2倍重量外信印刷物（10円＋5円）	double weight printed matter bearing 10 yen stamp and 5 yen stamp	13,000					

●**20 yen**　C291

地域 Area		郵便種別 Type of Mail	C291
国内郵便 domestic mail	1枚貼2倍重量書状	double weight letter bearing 20 yen stamp	10,000
	各種国内便	various domestic mail	2,000
外国郵便 international mail	1枚貼外信私製はがき	postcard bearing 20 yen stamp	15,000

●**24 yen**　C234 C277 C278

地域 Area		郵便種別 Type of Mail	C234	C277	C278
外国郵便 international mail	1枚貼外信書状	printed matter bearing 24 yen stamp	7,000	20,000	8,000
	各種外信便	various international mail	5,000	3,000	3,000

●**30 yen**　C292 C438 C416

地域 Area		郵便種別 Type of Mail	C292	C438	C416
国内郵便 domestic mail	1枚貼速達料金	express fee bearing 30 yen stamp	3,000	4,000	7,000
	1枚加貼速達私製はがき	express postcard additionally bearing 30 yen stamp	2,500		
	3倍重量書状	triple weight letter bearing 30 yen stamp		5,000	10,000
	各種国内便	various domestic mail	2,000	2,000	
外国郵便 international mail	1枚貼外信書状	printed matter bearing 30 yen stamp	4,000		
	第1地帯宛航空便印刷物	air mail printed matter to first zone	3,000		
	第2地帯宛航空便印刷物	air mail printed matter to second zone	3,000		
	第1地帯宛航空便私製はがき	air mail postcard to first zone		7,000	12,000
	第1地帯宛航空便印刷物	air mail printed matter to first zone		7,000	12,000
	各種外信便	various international mail			3,000

●**40 yen**　C429 C417

地域 Area		郵便種別 Type of Mail	C429, C417
国内郵便 domestic mail	1枚貼速達書状	express letter bearing 40 yen stamp	1,500
	1枚加貼書留書状	registered letter additional bearing 40 yen stamp	1,500
	各種国内便	various domestic mail	2,000
外国郵便 international mail	1枚貼外信書状	letter bearing 40 yen stamp	3,000
	第2地帯宛航空便私製はがき	air mail postcard to second zone	3,000
	第2地帯宛航空便2枚貼書状	air mail letter bearing two 40 yen stamps to second zoon	3,000
	第2地帯宛航空便印刷物	air mail printed matter to second zone	3,000

●**50 yen**　C418

地域 Area		郵便種別 Type of Mail	C418
国内郵便 domestic mail	1枚貼書留書状	registered letter bearing 50 yen stamp	2,000
外国郵便 international mail	第1地帯宛航空便書状	air mail letter to first zone	4,000
	第3地帯宛航空便私製はがき	air mail postcard to third zone	4,000
	第3地帯宛航空便2枚貼書状	air mail letter bearing two 50 yen stamps to third zone	4,000
	第3地帯宛航空便印刷物	air mail printed matter to third zone	4,000
	航空書簡	aerogramme bearing 50 yen stamp	4,000
	各種外信便	various international mail	2,000

●**CP1 切手帳 booklet pane（10 yen × 10）**

地域 Area		郵便種別 Type of Mail	CP1
国内郵便 domestic mail	タブ付1枚貼書状（8番切手）	letter bearing 10 yen stamp（pos.8）with tab	40,000
	タブ付1枚貼書状（1番切手）	letter bearing 10 yen stamp（pos.1）with tab	35,000
	タブ付1枚貼書状（11番切手）	letter bearing 10 yen stamp（pos.11）with tab	40,000
	タブ付1枚貼書状（6番切手）	letter bearing 10 yen stamp（pos.6）with tab	35,000

国民体育大会 National Athletic Meet

● 5 yen　A：第7回〜第9回 7th - 9th（C230-231, C239-240, C245-246）　B：第10回〜第15回 10th - 15th（C250-251, C258-259, C267-268, C282-283, C301-302, C322-323）　C：第16回〜第20回 16th - 20th（C344-345, C383-384, C401-402, C409-410, C436-437）

地域 Area		郵便種別 Type of Mail	A	B	C
国内郵便 domestic mail	1枚貼私製はがき	postcard bearing 5 yen stamp	2,500	1,000	700
	連刷ペア貼書状	letter bearing se-tenant of 5 yen stamps	3,500	1,500	1,200
	連刷ペア貼印刷物	printed matter bearing se-tenant of 5 yen stamps	3,000	1,500	1,200
	各種国内便	various domestic mail	3,000	1,500	1,200
外国郵便 international mail	連刷ペア貼外信印刷物	printed matter bearing se-tenant of 5 yen stamps	4,500	2,500	
	連刷3枚ストリップ貼2倍重量外信印刷物	double weight printed matter bearing strip of three of 5 yen stamps	8,000	6,000	
	3枚ストリップ貼外信印刷物	printed matter bearing strip of three of 5 yen stamps			4,000

切手趣味週間 Philatelic Week

● 10 yen　（C252, C260, C269, C340）　A：C294, C310, C378　B：C273, C387, C407, C425

地域 Area		郵便種別 Type of Mail	C252, C260	C269, C340	A	B
国内郵便 domestic mail	1枚貼書状	letter bearing 10 yen stamp	4,000	2,000	2,500	1,000
	2枚貼2倍重量書状	double weight letter bearing two 10 yen stamps	7,000	3,000	3,500	2,000
	1枚貼印刷物	printed matter bearing 10 yen stamp	-	-	2,500	1,500
	各種国内便	various domestic mail	4,000	2,000	3,000	1,500
外国郵便 international mail	1枚貼外信印刷物	printed matter bearing 10 yen stamp	6,000	3,500	4,000	2,000

国際文通週間 International Letter Writing Week

● 24 yen　C281

地域 Area		郵便種別 Type of Mail	C281
外国郵便 international mail	1枚貼外信書状	printed matter bearing 24 yen stamp	10,000
	各種外信便	various international mail	3,000

● 30 yen　C229　C320　C346

地域 Area		郵便種別 Type of Mail	C229	C320	C346
国内郵便 domestic mail	1枚貼速達私製はがき	express postcard bearing 30 yen stamp	12,000	15,000	8,000
	1枚加貼速達料金	express fee additionally bearing 30 yen stamp	8,000	10,000	
	1枚貼3倍重量書状	triple weight letter bearing 30 yen stamp	13,000		12,000
	各種国内便	various domestic mail	2,000	2,000	
外国郵便 international mail	1枚貼外信書状	printed matter bearing 30 yen stamp	10,000	13,000	
	第1地帯宛航空便印刷物	air mail printed matter to first zone	8,000	10,000	10,000
	第2地帯宛航空便印刷物	air mail printed matter to second zone	8,000	10,000	10,000
	各種外信便	various international mail	3,000	4,000	3,000

● 40 yen　A：C382　C339　C422　C434

地域 Area		郵便種別 Type of Mail	A
国内郵便 domestic mail	1枚貼速達料金	express fee bearing 40 yen stamp	6,000
	1枚加貼留料金	registered fee additionally bearing 40 yen stamp	5,000
外国郵便 international mail	1枚貼外信書状	printed matter bearing 40 yen stamp	13,000
	第2地帯宛航空便印刷物	air mail printed matter to second zone	10,000
	第1地帯宛航空便私製はがき	air mail postcard to first zone	10,000
	各種外信便	various international mail	3,000

東京1964オリンピック競技大会 Tokyo 1964 Olympic Games

● 5 yen + 5 yen　A：第1次〜第2次 1st issue – 2nd issue（C347 - C352）　B：第3次〜第4次 3rd issue - 4th issue（C353 - C358）　C：第5次〜第6次 5th issue - 6th issue（C359 - C366）

地域 Area		郵便種別 Type of Mail	A	B	C
国内郵便 domestic mail	1枚貼私製はがき	postcard bearing 5 yen stamp	12,000	10,000	8,000
	2枚貼書状	letter bearing two 5 yen stamps	12,000	10,000	7,000
	3枚貼書状	letter bearing three 5 yen stamps	10,000	10,000	5,000
	2枚貼印刷物	printed matter bearing two 5 yen stamps	12,000	12,000	7,000
	各種国内便	various domestic mail	5,000	5,000	3,000
外国郵便 international mail	3枚貼外信印刷物	printed matter bearing three 5 yen stamps	12,000	12,000	7,000

● 小型シート souvenir sheet
D：第1次〜第4次 1st issue – 4th issue（C367 - C370　15 yen）
E：第5次〜第6次 5th issue - 6th issue（C371 - C372　20 yen）

地域 Area		郵便の種類 tyoe of mail	D	E
国内郵便 domestic mail	2倍重量書状	double weight letter bearing 20 yen souvenir sheet		25,000
外国郵便 international mail	外信印刷物	printed matter bearing 15 yen souvenir sheet	u	12,000

記念・特殊切手（使用例評価／消印別評価）

消印別評価　Valuations by Postmark

この項の評価は、切手の印面に消印が90％以上程度かかっている単片で、局名や年月日がわかるものを基準としている。局名の違いなどによる評価の違いは考慮していない。

The value in this section is based on postmarks of single stamps, for which the name of the post office and full date can be read as follows;Clear strikes of 90% image or more. Differences in value due to differences in post office names, etc. are not taken into account.

◆C225 - C438（国民体育大会、切手趣味週間、国際文通週間、東京1964オリンピック競技大会を除く　Excluding National Athletic Meet, Philatelic Week, International Letter Writing Week and Tokyo 1964 Olympic Games）

国内便速達書状
長野・小布施 昭和 41.6.9
料金 40円（書状 10円+速達 30円）
適用期間：昭和 41.6.1 ～ 41.6.30
（昭和 41.7.1 ～ 書状 15円、速達 50円）
domestic express envelope / Obuse, June 9, 1966 / postal rate: 40 yen (envelope 10 yen + express fee 30 yen) / rate application period: June 1, 1966 - June 30, 1966

和文機械印（トビ色）
東京中央　昭和 28.10.13
適用期間：昭和 28.10.12 ～ 28.10.14
（昭和 28.10.15 ～ 黒色に変更）
domestic postcard, reddish brown machine datestamp / Tokyo C.P.O., October 13, 1953 / rate application period: October 12. 1953 - October 14, 1953

カタログ番号 Catalogue No.	額面 Face Values	消印タイプ Postmarks	和文印 Postmarks for Domestic Mail				欧文印 Postmarks for International Mail			
			櫛型印 JCD	鉄道郵便印 JCD-RW	機械印 JMD	ローラー印 JRD	櫛型印 RCD	三日月型印 RSD	機械印 RMD	ローラー印 RRD
									図1 fig.1	図2 fig.2
C225	5 yen		1,000	—	2,500	—	—	—	—	—
C226	10 yen		1,300	—	3,000	—	—	—	—	—
C227	5 yen		1,000	2,500	2,500	1,000	7,000	2,500	—	—
C228	10 yen		1,300	2,500	3,000	1,500	8,000	2,500	—	—
C229	10 yen		1,300	2,500	5,000	1,500	8,000	2,500	—	—
C232	5 yen		500	2,500	2,000	1,300	10,000	2,500	—	—
C233	10 yen		700	—	—	1,300	10,000	2,500	—	—
C234	24 yeb		2,500	—	—	4,000	13,000	4,000	—	10,000
C236	10 yen		1,000	—	—	1,300	—	2,500	—	—
C237	5 yen		800	—	5,000	1,000	—	2,000	—	—
C238	10 yen		1,000	—	5,000	1,300	—	2,500	—	—
C241 - C249	10 yen		1,000	2,500	5,000	1,300	—	2,000	—	—
C247	5 yen		800	2,500	1,300	1,000	—	2,000	—	—
C253 - C260	10 yen		500	2,000	3,000	800	—	1,500	10,000	3,000
C255	5 yen		300	1,500	700	500	—	1,300	—	2,500
C261 - C262	10 yen		800	2,000	3,000	1,200	—	2,000	10,000	—
C263 - C271	10 yen		300	2,000	3,000	500	—	1,300	—	—
C272 - C304	10 yen		200	1,000	2,000	300	—	1,000	8,000	2,000
C275	5 yen		200	—	500	300	—	1,000	—	—
C276	10 yen		200	—	3,000	300	—	1,000	—	—
C277	14 yen		300	—	—	500	—	1,500	—	2,500
C278	24 yen		300	—	—	500	—	1,500	—	—
C289	5 yen		200	1,000	—	300	—	1,000	—	—
C290	10 yen		200	—	—	300	—	1,000	—	—
C291	20 yen		500	—	—	700	—	1,500	—	—
C292	30 yen		500	3,000	—	700	—	1,500	10,000	3,000
C305 - C306	10 yen		200	—	—	500	—	1,000	—	—
C307 - C309	10 yen		300	2,000	3,000	500	—	1,200	8,000	3,000
C311	10 yen		200	—	—	300	—	1,000	—	—
C312	30 yen		400	—	—	500	—	1,500	10,000	—
C314 - C326	10 yen		200	1,000	—	300	—	1,000	8,000	—
C318	5 yen		200	—	—	300	—	1,000	—	—
C325	5 yen		200	—	—	300	—	1,000	—	—
C327 - C338	10 yen		500	—	4,000	800	—	1,500	—	3,000
C339	10 yen		500	—	4,000	1,000	—	2,000	—	—
C341 - C373	10 yen		200	—	—	300	—	1,000	—	—
C374 - C440	10 yen		300	1,000	1,000	500	—	1,000	7,000	3,000
C429	40 yen		500	—	1,500	500	—	1,500	—	—
C438	30 yen		500	—	1,500	500	—	1,500	—	2,500

図1 fig.1　欧文機械印 RMD

図2 fig.2　欧文ローラー印 RRD

凡例　Legend

Datestamp Type
<For Domestic Mail>
JCD: Comb Type Datestamps
JMD: Machine Datestamps
JRD: Roller Datestamps
<for International Mail>
RCD: Comb Type Datestamps
RMD: Machine Datestamps
RSD: Swordguard Type Datestamps
RRD: Roller Datestamps

Postmark Type
NY: New Year postmarks
RW: railway post office postmarks

記念・特殊切手（消印別評価）

◆国民体育大会 National Athletic Meet

カタログ番号 Catalogue No.	額面 Face Values	和文印 Postmarks for Domestic Mail					欧文印 Postmarks for International Mail			
		櫛型印 JCD	鉄道郵便印 JCD-RW	櫛型年賀印 JCD-NY	機械印 JMD	ローラー印 JRD	櫛型印 RCD	三日月型印 RSD	機械印 RMD	ローラー印 RRD
C230, C231	5 yen	1,000	3,000	8,000	2,000	1,500	8,000	2,500	—	5,000
C239, C240	5 yen	1,000	3,000	8,000	2,000	1,300	8,000	2,000	—	4,000
C245, C246	5 yen	800	3,000	7,000	2,000	1,300	10,000	2,000	—	4,000
C250, C251	5 yen	700	3,000	7,000	1,000	1,000	10,000	1,500	—	4,000
C258, C259	5 yen	500	1,500	—	900	900	14,000	1,500	—	4,000
C267, C268	5 yen	300	1,500	—	900	900	—	1,300	—	2,500
C282, C283	5 yen	300	1,500	—	800	800	—	1,300	12,000	2,500
C301, C302	5 yen	300	1,300	—	700	600	—	1,300	—	2,500
C322, C323	5 yen	300	1,300	—	700	600	—	1,300	—	2,500
C344, C345	5 yen	300	1,000	8,000	700	500	—	1,000	—	2,500
C383, C384	5 yen	200	1,000	—	700	500	—	1,000	—	2,000
C401, C402	5 yen	200	—	7,000	600	400	—	1,000	—	2,000
C409, C410	5 yen	200	—	—	600	400	—	800	10,000	2,000
C436, C437	5 yen	200	—	—	500	400	—	800	—	—

◆切手趣味週間 Philatelic Week

カタログ番号 Catalogue No.	額面 Face Values	和文印 Postmarks for Domestic Mail				欧文印 Postmarks for International Mail		
		櫛型印 JCD	鉄道郵便印 JCD-RW	機械印 JMD	ローラー印 JRD	三日月型印 RSD	機械印 RMD	ローラー印 RRD
C252	10 yen	2,500	6,000	8,000	3,000	3,500	—	—
C260	10 yen	2,500	6,000	8,000	3,000	3,500	—	—
C269	10 yen	1,000	—	—	1,000	1,500	—	—
C273	10 yen	300	1,500	—	300	500	—	1,500
C294	10 yen	600	2,000	4,000	600	1,000	—	—
C310	10 yen	800	—	—	1,000	1,200	—	—
C340	10 yen	700	—	—	1,000	1,100	—	3,000
C378	10 yen	700	—	2,000	1,000	1,100	—	—
C387	10 yen	400	—	2,000	500	1,000	—	—
C407	10 yen	300	—	2,000	300	500	—	—
C425	10 yen	200	—	2,000	200	400	—	—

◆国際文通週間 International Letter Writing Week

カタログ番号	額面	JCD	JCD-RW	JMD	JRD	RSD	RMD	RRD
C281	24 yen	700	—	—	4,000	3,000	—	—
C299	30 yen	700	—	10,000	4,000	3,000	—	—
C320	30 yen	1,200	—	—	—	5,000	—	8,000
C346	30 yen	1,500	—	—	2,000	5,000	10,000	—
C382	40 yen	1,300	—	—	—	5,000	—	8,000
C399	40 yen	800	—	—	800	5,000	—	—
C422	40 yen	600	—	—	800	5,000	—	—
C434	40 yen	600	—	—	800	5,000	—	—

◆東京1964オリンピック競技大会 Tokyo 1964 Olympic Games

カタログ番号	額面	JCD	JCD-RW	JMD	JRD	RSD	RMD	RRD
C347-349	5 yen + 5 yen	3,000	8,000	7,000	4,000	5,000	—	—
C350-352	5 yen + 5 yen	3,000	8,000	7,000	4,000	5,000	—	—
C353-355	5 yen + 5 yen	2,500	—	7,000	4,000	5,000	—	—
C356-358	5 yen + 5 yen	2,000	—	6,000	3,000	5,000	—	8,000
C359-362	5 yen + 5 yen	1,500	—	5,000	2,500	4,000	15,000	6,000
C363-366	5 yen + 5 yen	1,500	—	5,000	2,500	4,000	—	6,000
C414	5 yen	200	—	300	300	1,000	—	2,000
C415	10 yen	200	1,000	1,000	300	1,000	6,000	2,000
C416	30 yen	600	—	—	700	2,000	10,000	4,000
C417	40 yen	400	—	—	500	1,500	—	3,000
C418	50 yen	400	—	—	500	1,500	—	3,000

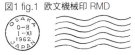

図1 fig.1 欧文機械印 RMD

図2 fig.2 欧文ローラー印 RRD

図3 fig.3 櫛型年賀印 JCD-NY

 C230-C251

 C258-C259

 C267-C437

1) JCD、JRD、RSD、RRDの東京オリンピック村局消印は、15,000円加算。
1) JCD, JRD, RSD and RRD postmarks of the Tokyo Olympic Village Post Office add 15,000 yen to the respective values in the table.

公園切手 National Park Stamps 1958-1966

■国立公園切手の発行

国立公園切手は、1931年（昭和6）制定の国立公園法により全国各地に国立公園が指定されたのを記念して発行された。

第1次国立公園切手（P1-90）は、1936年（昭和11）7月から1956年（昭31）10月まで発行された。第2次国立公園切手は、第1次国立公園切手発行終了6年後の1962年（昭和37）1月から1974年（昭和49）3月まで全24件52種（P91-142）が発行された。本カタログでは第2次国立公園切手の中から、書状10円時期に発行された12件28種（P91-118）を採録している。

■第2次国立公園切手の特徴

第2次国立公園切手は初期の富士箱根伊豆（P91-94）と日光（P95-98）は5円2種と10円2種の4種セットで、それ以降は5円、10円各1種ずつの2種セットで発行された。第1次国立公園切手とは異なり、小型シートは発行されなかった。

切手の原画にはすべて実際の写真が採用され、印刷は単色刷ではあるが、すべて多色刷グラビア輪転機で行われた。製版・目打組合せ型式は最初の富士箱根伊豆だけが2×2面順並び・4面穿孔（タイプ1C-1）で、それ以降は4面横または縦並び・連続穿孔（タイプ2B-1、2B-2）である。連続目打は、すべて二連1型である。

●公園名の文字の太さ（P106）
Thickness of Character "国立公園" for P106

正常 normal character

字が太い thick character

■国定公園切手の発行

国定公園切手は1958年（昭和33）8月から1973年（昭和48）3月まで22年間にわたり、全41件59種（P200-258）が発行された。本カタログではそのうち、書状10円時期に発行された22件26種（P200-225）を採録している。

■Issuance of the National Park Stamps

The National Park stamps were issued to commemorate the designation of National Parks throughout the country under the National Parks Law enacted in 1931 (Showa 6).

The first series of the National Park stamps (P1 - P90) was issued from July 1936 (Showa 11) to October 1956 (Showa 31). Six years after the completion of the first National Park Series, the second series was issued from January 1962 (Showa 37) to March 1974 (Showa 49). The total number of stamps in the second National Park series was 52, representing 24 National Parks (P91 - P142). This catalogue includes 28 stamps representing 12 National Parks (P91 - P118) from the second National Parks series, issued during the 10 yen letter-rate period.

■Features of the second National Park series stamps

The first and second sets, for Fuji-Hakone-Izu (P91 - P94) and Nikko (P95 - P98), of the second National Park series consisted of four stamps each: two 5 yen stamps and two 10 yen stamps. Two stamps each, one 5 yen stamp and one 10 yen stamp, were issued for each National Park thereafter. Unlike the first series, souvenir sheets were not issued for the second National Park series.

Actual photographs were used for all designs of the second National Park stamps series. Although all the designs are monochrome, the stamps were printed on the multi-color rotary photogravure press. The configuration of plate and perforation was type 1C-1 for Fuji-Hakone-Izu National Park only, and after that it was type 2B-1 or type 2B-2. The perforation type for the second National Park series stamps was the continuous tandem comb perforation with one extension hole.

■Issuance of the Quasi-National Park Stamps

The Quasi-National Park issue consisted of 59 stamps, representing 41 Quasi-National Parks, issued from August 1958 (Showa 33) to March 1973 (Showa 48). Among these stamps, this catalogue includes 26 stamps, representing 22 Quasi-National Parks, issued during the 10 yen letter rate period.

■国定公園切手の特徴

　国定公園切手の原画はすべて手描きであり、グラビア多色刷りが特徴である。採録範囲では初期の3件と最後の室戸阿南海岸（P224-225）が2種セット、その他は1種のみの発行で、額面はすべて10円である。

　印刷はすべてグラビア多色刷輪転機で行われ、製版・目打組合せ型式は最初の佐渡弥彦（P200-201）から琵琶湖（p210）までと大沼（P212）が順並び2×2面順合せ・4面穿孔（タイプ1C-1）で、山陰海岸（P211）と北長門海岸（P213）以降は4面横または縦並び・連続穿孔（タイプ2B-1、2B-2）である。連続目打は水郷（P216）までは二連2型であったが、金剛生駒（P215）では二連1型も併用され、石鎚（P217）以降は二連1型だけとなった。国定公園切手では、目打穿孔方式の意図的な試行が行われたと考えられる。また、玄海（P218）の右耳紙、伊豆七島（P219）の一部のシートで左耳紙に変則目打が存在する。

　なお、第2次国立公園切手、国定公園切手ではスクリーン角度の違いは確認されていない。

■Features of the Quasi-National Park Stamps

The designs of the Quasi-National Park stamps were all based on original art, and all are printed in multi-color photogravure. Two stamps each were issued for the first three Quasi-National Parks in the series, and for the last (Muroto-Anan-Kaigan, P224 - P225); one stamp each was issued for all the other Quasi- National Parks. All the stamps had a face value of 10 yen.

The multi-color rotary photogravure press was used to print these stamps. Configurations of plate and perforation were as follows; type 1C-1: P200, P201, P210, P212, types 2B-1 and 2B-2: P211, P213-P225. Types of continuous comb perforations were as follows; double comb with two extension holes: P200-P216, double comb with one extension hole: P217-P225. For P215, both one extension hole and two extension holes were used. Irregular perforations exist for P218（right selvage）and P219（left selvage）.

No screen angle varieties have been reported for both the second National Park series stamps and the Quasi-National Park stamps.

●目打穿孔（P215）Perforation Method for P215

（左）二連2型 (left) comb perforation with two extension holes

（右）二連1型 (right) comb perforation with one extension hole

●切手原画（P219）Original Art for P219

原画作者:木村 勝　designer:Kimura Masaru

共通切手データ Common Information

版式：グラビア輪転版	printing plate : rotary photogravure plate
目打：櫛型13½	perforation : comb perf. 13½
印面寸法：横型33×22.5mm、縦型：22.5×33mm	printing area size : horizontal = 33×22.5mm, vertical = 22.5×33mm
シート構成：横型20枚（横5×縦4）、縦型20枚（横4×縦5）	pane format : horizontal = 20 subjects of 5×4, vertical = 20 subjects of 4×5
銘版・銘版位置：⑬・19番	imprint : type ⑬ pos.19

◆原画作者 Designers

原画作者 Designers	カタログ番号 Catalogue No.
大塚均 Otsuka Hitoshi	P202, P203, P206, P209, P213, P216, P223
鴨田順三 Kamoda Zyunzo	P91, P97
木村勝 Kimura Masaru	P204, P205, P211, P212, P218, P219
斎藤栄嗣 Saito Eiji	P98, P101
丹後昭 Tago Akira	P93, P95
中島桂 Nakajima Kei	P92, P96, P99, P102

原画作者 Designers	カタログ番号 Catalogue No.
錦織健次郎 Nishikiori Kenjiro	P94
長谷部日出男 Hasebe Hideo	P200, P201, P207, P210, P217, P220, P222
久野實 Hisano Minoru	P208, P214, P221
星川忠道 Hoshino Tadamichi	P100
渡邊三郎 Watanabe Saburo	P215, P224, P225
不明 unknown	P103-P118

I. 第2次国立公園切手 2nd National Park Series 1962-1965

P91-94 富士箱根伊豆国立公園　発行日：1962（昭和37）.1.16
Fuji-Hakone-Izu National Park　Date of Issue : 1962 (Showa 37) .1.16

P91 芦ノ湖と富士
Mt. Fuji from Lake Ashi

P92 石廊崎・簑掛岩
Minokake-Iwa at Irozaki

P93 三ツ峠からの富士
Mt. Fuji from Mitsu Pass

P94 大瀬崎と富士
Mt.Fuji from Cape of Ose

●風景印 Pictorial Postmarks

 富士 Fuji

御殿場 Gotenba

土肥 Toi

 松崎 Matsuzaki

 三津 Mitsu

 箱根町 Hakone-machi

 河口 Kawaguchi

			**	●
P91	5 yen	こい緑 deep green	100	40
P92	5 yen	暗い青 dark blue	100	40
P93	10 yen	赤味茶 reddish brown	200	70
P94	10 yen	黒 black	200	70
	P91-94 (4種 set of 4 values)		600	220
	FDC (4種 set of 4 values)			1,500

切手データ Basic Information	
発行枚数：各800万枚	number issued : 8 million each
製版・目打組合せ：タイプ1C-1	plate and perforation configuration : type 1C-1

P91 ●定常変種 Constant Flaws

A-pos.1

A-pos.6

A-pos.12

B-pos.6

B-pos.18

C-pos.19

D-pos.8

D-pos.15

位置 position	特徴 feature	**	●
A-pos.1	湖に白抜け white dot on lake	200	100
A-pos.6	山頂に緑点 green dot on mountaintop		
A-pos.12	空に筋 streak in sky		
B-pos.6	"郵"の下方に小緑点 small green dot below "郵"		
B-pos.18	山頂の左に大きな白点 large white dot to left of mountaintop		
C-pos.19	"便"の左下に2連点 two dots to lower left of "便"		
D-pos.8	"郵"の下に緑点 green dot below "郵"		
D-pos.15	山頂の左に白点 white dot to left of mountaintop		

P92 ●定常変種 Constant Flaws

A-pos.10

B-pos.10　B-pos.2

C-pos.3

C-pos.19　D-pos.17　C-pos.11

位置 position	特徴 feature	**	●
A-pos.10	"便"の下に青点 blue dot below "便"	200	100
B-pos.2	白点2ヵ所 two white dots		
B-pos.10	"本"の下方に大きな白点 large white dot below "本"		
C-pos.3	空に小さな青点2ヵ所 two small blue dots in sky		
C-pos.11	"日"の下方に青点 blue dot below "日"		
C-pos.19	大きな白点 large white dot		
D-pos.17	空に白点 white dot in sky		

第2次国立公園切手

P93 ●定常変種 Constant Flaws

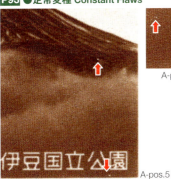
A-pos.5

A-pos.9

A-pos.11

A-pos.14

B-pos.5

C-pos.4

C-pos.13

D-pos.9

位置 position	特徴 feature	**	●
A-pos.5	山肌と"園"の下に白点2ヵ所 two white dots on mountainside and below "園"	400	200
A-pos.9	"富"の左上に白点 white dot to upper left of "富"		
A-pos.11	"1"の左に白点2個 two white dots to left of "1"		
A-pos.14	"郵"の下に茶点 brown dot below "郵"		
B-pos.5	"郵"と"便"の間に茶点 brown dot between "郵" and "便"		
C-pos.4	裾野の左に小茶点 small brown dot to left of mountain base		
C-pos.13	"本"と"郵"の間に茶点2個 two brown dots between "本" and "郵"		
D-pos.9	"日"の左下に茶点 brown dot to lower left of "日"		

P94 ●定常変種 Constant Flaws

A-pos.7

A-pos.13

B-pos.3

B-pos.12

C-pos.3

C-pos.9

D-pos.18

D-pos.10

位置 position	特徴 feature	**	●
A-pos.7	山頂の左に黒点 black dot to left of mountaintop	400	200
A-pos.13	"郵"の右上に黒点2個 two black dots to upper right of "郵"		
B-pos.3	"便"の下に黒点 black dot below "便"		
B-pos.12	黒点2ヵ所 two black dots		
C-pos.3	"本"の下に黒点 black dot below "本"		
C-pos.9	林の上に黒点 black dot above forest		
D-pos.10	白点2ヵ所 two white dots		
D-pos.18	L字形の白抜き L字形の白抜き white L-shaped line		

P95-98 日光国立公園 Nikko National Park

発行日：1962（昭和37）.9.1
Date of Issue：1962（Showa 37）.9.1

P95 尾瀬ヶ原と至仏山
Ozegahara Swampland and Mt. Sihbutsu

P96 那須茶臼岳と噴煙
Plume on Mt. Chausu, Nasu

P97 中禅寺湖の八丁出島と男体山 Lake Chuzenji and Mt. Nantai

P98 塩原渓谷潜竜峡
Senryu-kyo Narrows, Shiobara

●風景印 Pictorial Postmarks

P95-98 日光 Nikko

		**	●
P95	5 yen 緑味青 greenish blue	50	40
P96	5 yen 赤茶 maroon	50	40
P97	10 yen こい紫 purple	70	40
P98	10 yen 黄茶 olive	70	40
	P95-98（4種 set of 4 values）	240	160
FDC	（4種 set of 4 values）		1,000

切手データ Basic Information	
発行枚数：各900万枚	number issued：9 million each
製版・目打組合せ：タイプ2B-2	plate and perforation configuration：type 2B-2

1) この切手以降、国立公園切手は連続櫛型目打の二連1型で穿孔されるようになった。

1) The continuous comb perforation with single extension hole in tandem was used for these National Park Stamps and thereafter.

第２次国立公園切手

P95 ●定常変種 Constant Flaws

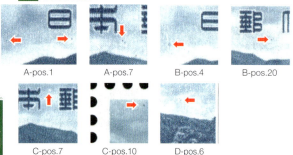

位置 position	特徴 feature	**	●
A-pos.1	青点2ヵ所 two blue dots	200	100
A-pos.7	"本"の右下に2連点 two dots to lower right of "本"		
B-pos.4	"日"の左下方に青点 blue dot to lower left of "日"		
B-pos.20	"郵"の右下に青点 blue dot to lower right of "郵"		
C-pos.7	"本"と"郵"の間に白点 white dot between "本" and "郵"		
C-pos.10	第1コーナーに青点 blue dot in NW corner		
D-pos.6	空に青の2連点 two blue dots in sky		

P96 ●定常変種 Constant Flaws

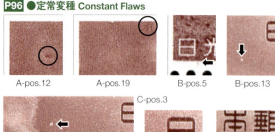

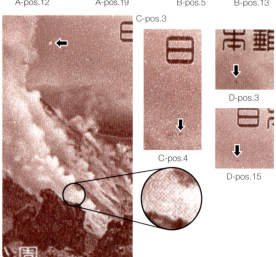

位置 position	特徴 feature	**	●
A-pos.12	噴煙の中に大きな茶点 large maroon dot in plume	200	100
A-pos.19	噴煙の中に濃いあざ dark bruises in plume		

右上につづく Continued to up right ↗

position	feature
B-pos.5	マージンに茶点 brown dot in bottom margin
B-pos.13	"日"の左下に大きな白点 large white dot to lower left of "日"
C-pos.3	白点2ヵ所 two white dots
C-pos.4	"日"の下方に茶点 maroon dot below "日"
D-pos.3	"本"の下方に大きな茶点 large maroon dot below "本"
D-pos.15	"日"の下方に茶点 maroon dot below "日"

P97 ●定常変種 Constant Flaws

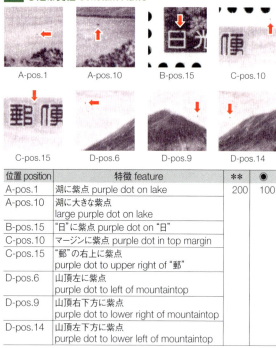

位置 position	特徴 feature	**	●
A-pos.1	湖に紫点 purple dot on lake	200	100
A-pos.10	湖に大きな紫点 large purple dot on lake		
B-pos.15	"日"に紫点 purple dot on "日"		
C-pos.10	マージンに紫点 purple dot in top margin		
C-pos.15	"郵"の右上に紫点 purple dot to upper right of "郵"		
D-pos.6	山頂左に紫点 purple dot to left of mountaintop		
D-pos.9	山頂右下方に紫点 purple dot to lower right of mountaintop		
D-pos.14	山頂左下方に紫点 purple dot to lower left of mountaintop		

P98 ●定常変種 Constant Flaws

位置 position	特徴 feature	**	●
A-pos.8	森の中に霞 mist in forest	200	100
B-pos.13	"便"の右上に黄茶点 olive dot to upper right of "便"		
C-pos.12	"日"に黄茶点 olive dot on "日"		
D-pos.4	川の中に大きな黄茶点 large olive dot in river		

P99-100 雲仙天草国立公園　発行日：1963（昭和38）.2.15
Unzen-Amakusa National Park　Date of Issue : 1963 (Showa 38) .2.15

P99 普賢岳の霧氷
Frost Flowers on Mt. Fugen

P100 天草松島と雲仙
Amakusa Islands and Mt. Unzen

●風景印
Pictorial Postmarks

P99-100
雲仙 Unzen

		**	●
P99	5 yen 暗い青 greyish blue	40	40
P100	10 yen 赤 carmine	50	40
	P99-100 (2種 set of 2 values)	90	80
FDC	(2種 set of 2 values)		700

切手データ Basic Information	
発行枚数：各1,000万枚	number issued : 10 million each
製版・目打組合せ：タイプ2B-2	plate and perforation configuration : type 2B-2

P99 ●定常変種 Constant Flaws

A-pos.1　A-pos.2　　B-pos.11

B-pos.19

C-pos.1　C-pos.12　D-pos.5　D-pos.19

位置 position	特徴 feature	**	●
A-pos.1	"郵"の左上に白点 white dot to upper left of "郵"	200	100
A-pos.2	"本"の中に青点 blue dot in "本"		
B-pos.11	"便"の右の雲中に青点 blue dot in cloud to right of "便"		
B-pos.19	上部3ヵ所に小青点 three small blue dots on top		
C-pos.1	空に青点 blue dot in sky		
C-pos.12	海に青ボート blue "boat" in ocean		
D-pos.5	空に青点 blue dot in sky		
D-pos.19	"刷"と"局"の間に小点 small dot between "刷" and "局"		

P100 ●定常変種 Constant Flaws

A-pos.10　　A-pos.19　　B-pos.4

B-pos.8

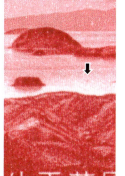
B-pos.5

B-pos.16

位置 position	特徴 feature	**	●
A-pos.10	右マージンに赤点 red dot in right margin	200	100
A-pos.19	マージンに赤点 red dot in top margin		
B-pos.4	島の上部に赤点 red dot above island		
B-pos.5	島々の間に赤点 red dot between islands		
B-pos.8	"郵"に淡い赤丸点 pale circular red dot on "郵"		
B-pos.16	"日"と島影の間に赤点 red dot between "日" and island		

第2次国立公園切手

P101-102 白山国立公園　発行日：1963（昭和38）.3.1
Hakusan National Park　Date of Issue：1963(Showa 38).3.1

P101 白山翠ヶ池
Midorigaike (Green Pond)

P102 白山連邦
Hakusan Range

●風景印
Pictorial Postmarks

P101-102
白峰 Shiramine

		**	●
P101	5 yen 茶黒 violet brown	40	40
P102	10 yen 暗い青緑 dark green	50	40
P101-102	(2種 set of 2 values)	90	80
FDC	(2種 set of 2 values)		500

切手データ Basic Information	
発行枚数：各1,000万枚	number issued : 10 million each
製版・目打組合せ：タイプ2B-2	plate and perforation configuration : type 2B-2

P101 ●定常変種 Constant Flaws

A-pos.5　　B-pos.1　　B-pos.13　　C-pos.3

C-pos.17　　　D-pos.11

位置 position	特徴 feature	**	●
A-pos.5	"便"の右上に茶点 brown dot to upper right of "便"	200	100
B-pos.1	"郵"の上方に茶点 brown dot above "郵"		
B-pos.13	"5"の右上に茶点 brown dot to upper right of "5"		
C-pos.3	池の上に濃い茶点 dark brown dot above pond		
C-pos.17	池の上に白抜け white dot above pond		
D-pos.11	池の中に大きな茶点 large brown dot in pond		

P102 ●定常変種 Constant Flaws

A-pos.3

A-pos.5

A-pos.14

B-pos.3

D-pos.5

C-pos.19

位置 position	特徴 feature	**	●
A-pos.3	"郵"と"便"の間に緑点と白点 green dot and white dot between "郵" and "便"	200	100
A-pos.5	枝の下に緑点と白点 green dot and white dot below branches		
A-pos.14	"日本"の下に白点 white dot below "日本"		
B-pos.3	"便"の下方に大きな白点 large white dot below "便"		
C-pos.19	山頂の上部に小点、"刷"の上に横線 small dot above mountaintop and line above "刷" of imprint		
D-pos.5	"本"の右下に緑点 green dot to lower right of "本"		

P103-104 磐梯朝日国立公園 Bandai-Asahi National Park

発行日：1963（昭和38）.5.25
Date of Issue : 1963 (Showa 38) .5.25

P103 以東岳
Ito-dake, Asahi Range

P104 檜原湖と磐梯山
Lake Hibara and Mt. Bandai

● 風景印
Pictorial Postmarks

P103-104
裏磐梯 Urabandai

		**	●
P103	5 yen 青緑 green	40	40
P104	10 yen 茶 reddish brown	50	40
	P103-104 (2種 set of 2 values)	90	80
FDC	(2種 set of 2 values)		450

切手データ Basic Information	
発行枚数：各1,100万枚	number issued : 11 million each
製版・目打組合せ：タイプ2B-2	plate and perforation configuration : type 2B-2

P103 ● 定常変種 Constant Flaws

A-pos.19

B-pos.19

B-pos.20

C-pos.5

D-pos.5

C-pos.20

D-pos.6

位置 position	特徴 feature	**	●
A-pos.19	"本"の右に緑点 green dot to right of "本"	200	100
B-pos.19	右上の雲に6個の緑点 six green dots above right cloud		
B-pos.20	右上の雲の上に濃い点群 dense dots in right cloud		
C-pos.5	右上の空に緑点と白抜け green dot and white dot in right sky		
C-pos.20	左上の雲に5個の緑点 five green dots in left cloud		
D-pos.5	雲の左に大きなベタ large sticky to left of cloud		
D-pos.6	雲の上に大きな緑点、雲の中に点 large green dot above cloud, dot in cloud		

P104 ● 定常変種 Constant Flaws

A-pos.3

A-pos.20

B-pos.5

B-pos.20

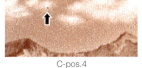
C-pos.4

C-pos.5

D-pos.6

D-pos.11

位置 position	特徴 feature	**	●
A-pos.3	雲の左に茶の塊 brown lump to left of cloud	200	100
A-pos.20	山頂の右上に茶の塊 brown lump to upper right of mountaintop		
B-pos.5	"1"の中に茶点 brown dot in "1"		
B-pos.20	"磐"の上に白点 white dot above "磐"		
C-pos.4	山頂に茶点 brown dot above mountaintop		
C-pos.5	雲の左に茶の塊 brown lump to left of cloud		
D-pos.6	"日"の右下方に茶の塊 brown block to lower right of "日"		
D-pos.11	"日"の右下方に大きな茶の塊 large brown block to lower right of "日"		

第2次国立公園切手

P105-106 瀬戸内海国立公園　発行日：1963（昭和38）.8.20
Setonaikai (Inland Sea) National Park　Date of Issue: 1963 (Showa 38) .8.20

P105 鷲羽山
View of Mt. Washu

P106 鳴門の渦潮
Whirlpool at Naruto

●風景印　Pictorial Postmarks

P105-106 鳴門 Naruto

		**	●
P105	5 yen 暗い黄茶 olive bistre	40	30
P106	10 yen 暗い青緑 dark green	50	30
	P105-106 (2種 set of 2 values)	90	60
	FDC (2種 set of 2 values)		400

切手データ Basic Information	
発行枚数：各1,400万枚	number issued: 14 million each
製版・目打組合せ：タイプ2B-2	plate and perforation configuration: type 2B-2

P105 ●定常変種 Constant Flaws

A-pos.13

A-pos.18

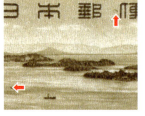

A-pos.19

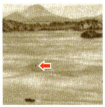
B-pos.10

C-pos.14

位置 position	特徴 feature	**	●
A-pos.13	"5"の上に茶点 bistre dot above "5"	200	100
A-pos.18	2ヵ所に大きな茶点 two large bistre dots		
A-pos.19	"5"の左上に茶点 bister dot to upper left of "5"		
B-pos.10	船の上に大きな茶点 large bistre dot above ship		
C-pos.14	島の上に白傷 white scar above island		

P106 ●定常変種 Constant Flaws

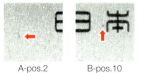
A-pos.2　B-pos.10　C-pos.6

D-pos.8

D-pos.3

正常 normal

A-D pos.18　A-D pos.19

位置 position	特徴 feature	**	●
A-pos.2	"日"の左下方に緑点 green dot to lower left of "日"	200	100
B-pos.10	"本"の左に緑点 green dot to left of "本"		
C-pos.6	山頂に緑点 green dot on mountaintop		
D-pos.3	山麓の上方に緑点 green dot above foot of mountain		
D-pos.8	"便"の下に緑点 green dot below "便"		
A-D pos.18	"園"の字が太い thick character "園"		
A-D pos.19	全部の字が太い thick characters "瀬戸内海国立公園"		

1) pos.18、pos.19にある公園名の文字転写バラエティは、シートA〜Dで共通している。原乾板作業中に発生したものと推測されるが、原因は不明である。

1) Why the features of characters for pos. 18 and pos. 19 are common to sheets A-D has not been clarified.

P107-108 大雪山国立公園　発行日：1963（昭和38）.9.1
Daisetsuzan National Park　Date of Issue：1963 (Showa 38).9.1

P107 然別湖
Lake Shikaribetsu

P108 層雲峡からの黒岳
Mt. Kurodake from Sounkyo Valley

●風景印
Pictorial Postmarks

P107-108
層雲峡 Sounkyo

		**	●
P107	5 yen こい緑味青 deep prussian blue	40	30
P108	10 yen こい紫 rose-violet	50	30
	P107-108 (2種 set of 2 values)	90	60
	FDC (2種 set of 2 values)		400

切手データ Basic Information	
発行枚数：各1,400s	number issued：14 million each
製版・目打組合せ：タイプ2B-2	plate and perforation configuration: type 2B-2

P107 ●定常変種 Constant Flaws

A-pos.9

A-pos.13

A-pos.18

B-pos.10

C-pos.18

D-pos.2

D-pos.14

D-pos.16

P108 ●定常変種 Constant Flaws

A-pos.3

A-pos.5

A-pos.11

B-pos.3

B-pos.19

C-pos.17

C-pos.19

D-pos.18

位置 position	特徴 feature	**	●
A-pos.9	"5"の上方に白抜け white spot above "5"	200	100
A-pos.13	山肌に白抜け white spot on mountainside		
A-pos.18	湖面に大きな白点 large white dot on lake		
B-pos.10	湖面に大きな青点 large blue dot on lake		
C-pos.18	湖面に大きな青点 large blue dot on lake		
D-pos.2	"便"の右下に青点 blue dot to lower right of "便"		
D-pos.14	マージンに青点 blue dot in top margin		
D-pos.16	"山"と"国"の下方に白点 white dot below "山" and "国"		

位置 position	特徴 feature	**	●
A-pos.3	マージンに小点2ヵ所 two small dots in bottom margin	200	100
A-pos.5	"郵"と"便"に紫点 purple dots beside "郵" and "便"		
A-pos.11	"日"の左に曲線 arc to left of "日"		
B-pos.3	山頂左に紫点 purple dot to left of mountaintop		
B-pos.19	"局"の右上に小点 small dot to upper right of "局" in imprint		
C-pos.17	"日"の下方に紫点 purple dot below "日"		
C-pos.19	"本"の上に紫点 purple dot above "本"		
D-pos.18	マージンに小点 small dot in top margin		

P109-110 伊勢志摩国立公園
Ise-Shima National Park

発行日：1964（昭和39）.3.15
Date of Issue：1964 (Showa 39) .3.15

P109 宇治橋
Uji Bridge

P110 鳥羽海岸
View of Toba Coast

●風景印 Pictorial Postmarks

P109-110
伊勢 Ise

P109 五十鈴川
Isuzugawa

P110 鳥羽
Toba

		**	◉
P109	5 yen 暗い茶 sepia	40	30
P110	10 yen こい赤紫 red-lilac	50	30
	P109-110 (2種 set of 2 values)	90	60
	FDC (2種 set of 2 values)		350

切手データ Basic Information

発行枚数：各1,750万枚	number issued：17.5 million each
製版・目打組合せ：タイプ2B-2	plate and perforation configuration：type 2B-2

P109 ●定常変種 Constant Flaws

A-pos.19

A-pos.20

B-pos.8

B-pos.14

C-pos.9

C-pos.19

D-pos.14

位置 position	特徴 feature	**	◉
A-pos.19	植込みの下方に小点 small dot below shrubbery	200	100
A-pos.20	"日"の右に茶点 sepia dot to right of "日"		
B-pos.8	橋の欄干に小点 small dot on bridge railing		
B-pos.14	川原に小点　small dot on riverbank		
C-pos.9	"郵"の下方に2つの茶点 two sepia dots below "郵"		
C-pos.19	木の横に茶点 sepia dot next to tree		
D-pos.14	橋の欄干の上方に茶点 sepia dot above bridge railing		

P110 ●定常変種 Constant Flaws

A-pos.16

B-pos.7

B-pos.15

B-pos.19

C-pos.1

D-pos.1

位置 position	特徴 feature	**	◉
A-pos.16	小点2ヵ所 two small dots	200	100
B-pos.7	"本"の右上に点 dot to upper right of "本"		
B-pos.15	"日"と"本"の間に点 dot between "日" and "本"		
B-pos.19	海面に濃い点 dark dot on sea		
C-pos.1	マージンに2つの点 two dots in top margin		
D-pos.1	"郵"と"便"の間に斜線 line between "郵" and "便"		

P111-112 大山隠岐国立公園 Daisen-Oki National Park
発行日：1965（昭和40）.1.20　Date of Issue : 1965 (Showa 40) .1.20

第2次国立公園切手

P111 赤松の池と大山
Mt. Daisen

P112 隠岐浄土ヶ浦
Paradise Cove, Oki Islands

●風景印 Pictorial Postmarks

P111 大山
Daisen

P112 西郷
Saigo

			**	●
P111	5 yen 灰味青 dark blue		40	30
P112	10 yen 赤味だいだい brown-orange		50	30
	P111-112 (2種 set of 2 values)		90	60
	FDC (2種 set of 2 values)			350

切手データ Basic Information	
発行枚数：P111-2,500万枚、P112-2,400万枚	number issued : P111=25 million, P112=24 million
製版・目打組合せ：タイプ2B-2	plate and perforation configuration : type 2B-2

1) この切手以降、国立公園切手のスクリーン線数は250線となった。
1) Following this issue, 250-line screens were used for the National Park Stamps.

P111 ●定常変種 Constant Flaws

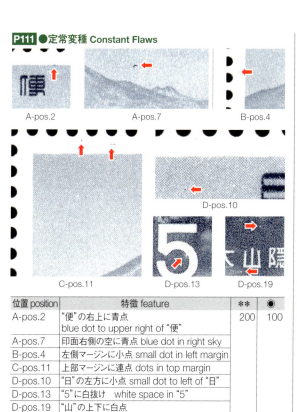

P112 ●定常変種 Constant Flaws

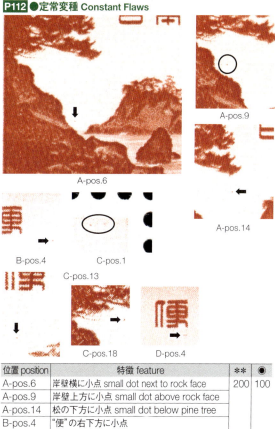

位置 position	特徴 feature	**	●
A-pos.2	"便"の右上に青点 blue dot to upper right of "便"	200	100
A-pos.7	印面右側の空に青点 blue dot in right sky		
B-pos.4	左側マージンに小点 small dot in left margin		
C-pos.11	上部マージンに連点 dots in top margin		
D-pos.10	"日"の左右に小点 small dot to left of "日"		
D-pos.13	"5"に白抜け white space in "5"		
D-pos.19	"山"の上下に白点 white dots above and below "山"		

位置 position	特徴 feature	**	●
A-pos.6	岸壁横に小点 small dot next to rock face	200	100
A-pos.9	岸壁上方に小点 small dot above rock face		
A-pos.14	松の下方に小点 small dot below pine tree		
B-pos.4	"便"の右下方に小点 small dot to lower right of "便"		
C-pos.1	第2コーナーに2つの点 two dots in NE corner		
C-pos.13	"便"の下方に小点 small dot below "便"		
C-pos.18	松の下方に小点 small dot below pine tree		
D-pos.4	"便"の下に2つの点 two dots below "便"		

P113-114 上信越高原国立公園
Jo-Shin-Etsu Kogen National Park
発行日：1965（昭和40）.3.15　Date of Issue：1965 (Showa 40) .3.15

P113 清津峡 Kiyotsu Gorge

P114 野尻湖と妙高山
Lake Nojiri and Mt. Myoko

● 風景印
Pictorial Postmarks

P113-114 野尻湖 Nojiriko

	**	●
P113　5 yen 暗い茶 brown	40	30
P114　10 yen 暗い赤紫 magenta	50	30
P113-114 (2種 set of 2 values)	90	60
FDC (2種 set of 2 values)		350

切手データ Basic Information	
発行枚数：P113-2,500万枚、P114-2,400万枚	number issued：P113=25 million, P114=24 million
製版・目打組合せ：P113-タイプ2B-1、P114-タイプ2B-2	plate and perforation configuration：P113=type 2B-1, P114=type 2B-2

P113 ● 定常変種 Constant Flaws

A-pos.12

B-pos.1

B-pos.2

C-pos.6

D-pos.1

位置 position	特徴 feature	**	●
A-pos.12	"5"の右上方に白点 white dot to upper right of "5"	200	100
B-pos.1	"本"の左下方に極小点 very small dot to lower left of "本"		
B-pos.2	"本"の左下方に小点 small dot to lower left of "本"		
C-pos.6	"5"の右に白点 white dot to right of "5"		
D-pos.1	"上"の左上方に白点 white dot to upper left of "上"		

P114 ● 定常変種 Constant Flaws

A-pos.3

A-pos.6

A-pos.17

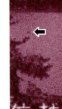

B-pos.9

B-pos.10

B-pos.13

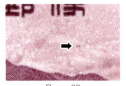
B-pos.20

C-pos.5

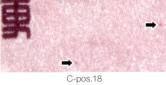
C-pos.18

D-pos.16　D-pos.19

B-pos.6

位置 position	特徴 feature	**	●
A-pos.3	"本"の左に小点 small dot to left of "本"	200	100
A-pos.6	左側マージンに小点 small dot in left margin		
A-pos.17	"日"の左に小点 small dot to left of "日"		
B-pos.6	枝の上方に濃い点 dark dot above branch		
B-pos.9	"便"の右に2つの点 two dots to right of "便"		
B-pos.10	左側山麓の上方に点 dot above foot of mountain		
B-pos.13	第2コーナーに濃い点 dark spot in NE corner		
B-pos.20	"郵"の下方に2つの点 two dots below "郵"		
C-pos.5	枝の上方に点 dot above branch		
C-pos.18	"便"の右方に2つの点 two dots to right of "便"		
D-pos.16	"高"の上方に白点 white dot above "高"		
D-pos.19	印面右辺に濃い点 dark dot beside right edge of printing area		

P115-116 阿蘇国立公園 / Aso National Park

発行日：1965（昭和40）.6.15
Date of Issue：1965 (Showa 40) .6.15

P115 中岳噴火口
Crater of Mt. Naka

P116 城山からの阿蘇五岳
Five Central Peaks of Aso,
view from Mt. Shiroyama

● 風景印 Pictorial Postmarks

P115-116 阿蘇
Aso

P115-116 赤水
Akamizu

		**	●
P115	5 yen こい紅赤 carmine rose	40	30
P116	10 yen こい緑 deep green	50	30
P115-116	(2種 set of 2 values)	90	60
FDC	(2種 set of 2 values)		350

切手データ Basic Information	
発行枚数：P115-2,500万枚、P116-2,400万枚	number issued：P115=25 million, P116=24 million
製版・目打組合せ：タイプ2B-2	plate and perforation configuration：type 2B-2

位置 position	特徴 feature
B-pos.10	右下マージンに赤点群 red dots in lower right margin
B-pos.15	左側噴煙に点 dot in left plume
C-pos.15	噴煙とマージンに点 dots in plume and top margin
D-pos.3	噴煙に濃い点と小点 dark dot and small dot in plume
D-pos.8	"便"の右に斜線 line to right of "便"
DD-pos.15/20	右側マージンに赤点 red dot in right margin

P115 ● 定常変種 Constant Flaws

A-pos.3

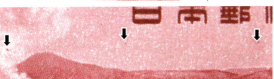
A-pos.4

B-pos.10　　　B-pos.15

C-pos.15

D-pos.3　　D-pos.8　　D-pos.15/20

位置 position	特徴 feature	**	●
A-pos.3	山麓に濃い点 dark dot above foot of mountain	200	100
A-pos.4	噴煙と空に3つの点 three dots on plume and sky		

右上につづく Continued to up right ↗

P116 ● 定常変種 Constant Flaws

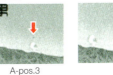
A-pos.3　　　A-pos.12

B-pos.5

B-pos.15

C-pos.2

C-pos.8

D-pos.6

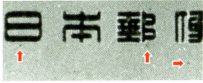
D-pos.16

位置 position	特徴 feature	**	●
A-pos.3	噴煙の上方に緑点 green dot above plume	200	100
A-pos.12	噴煙の右上方に緑点2個 two green dots to upper right of plume		
B-pos.5	右側山麓に点 dot above right foot of mountain		
B-pos.15	右側マージンに点 dot in right margin		
C-pos.2	"便"の左下に濃い緑点 dark green dot lower left of "便"		
C-pos.8	"便"の下に緑点、左下方に大きな白ぬけ Green dot below "便", large white spot to lower left of "便"		
D-pos.16	"日本郵便"の下に3つの点 three dots below "日本郵便"		

第2次国立公園切手

P117-118　知床国立公園　発行日：1965（昭和40）.11.15
Shiretoko National Park　1965(Showa 40) .11.15

P117
斜里海岸と
知床硫黄山
Mt. Iwo from
Shari Coast

P118 羅臼湖畔と羅臼岳
Rausu Lake and Mt. Rausu

●風景印 Pictorial Postmarks

P117 斜里　　P118 羅臼
Syari　　　　Rausu

			**	●
P117	5 yen 青緑 prussian green		40	30
P118	10 yen こい紫味青 bright blue		50	30
	P117-118 (2種 set of 2 values)		90	60
	FDC (2種 set of 2 values)			350

切手データ Basic Information	
発行枚数：P117-2,550万枚、P118-2,500万枚	number issued：P117=25.5 million, P118=25 million
製版・目打組合せ：P117-タイプ2B-1、P118-タイプ2B-2	plate and perforation configuration：P117=type 2B-1, P118=type 2B-2

P117 ●定常変種 Constant Flaws

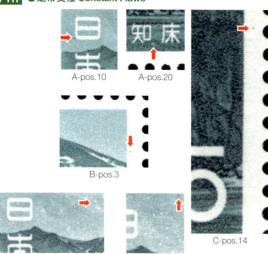

位置 position	特徴 feature	**	●
A-pos.10	"日"の左下に緑点 dot at lower left of "日"	200	100
A-pos.20	"知"の右下に白抜け white spot to lower right of "知"		
B-pos.3	右マージンに緑点 green dot in right margin		
C-pos.14	右マージンに緑点 green dot in right margin		
D-pos.3	空の中間に緑点 green dot in middle of sky		
D-pos.4	"日"の右上方に緑点 green dot to upper right of "日"		

P118 ●定常変種 Constant Flaws

　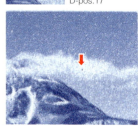

位置 position	特徴 feature	**	●
A-pos.11	湖に大きな白い玉 large white spot in lake	200	100
A-pos.20	"1"の下に白点 white dot below "1"		
B-pos.2	山頂の左上方に白点 white dot to upper left of mountaintop		
B-pos.15	山頂の右上方に白点 white dot to upper right of mountaintop		
C-pos.2	湖面に大きな濃い青点 large dark dot on lake		
D-pos.5	"郵"の下方に青点 blue dot below "郵"		
D-pos.17	雲の中に濃青点 dark blue dot in cloud		
D-pos.20	雲に小さな青点 small blue dot in cloud		

II. 国定公園切手 Quasi-National Park Series 1958-1966

P200-201 佐渡弥彦国定公園
Sado-Yahiko Quasi-National Park

発行日：1958（昭和33）.8.20
Date of Issue : 1958 (Showa 33).8.20

●風景印 Pictorial Postmarks

P200 外海府海岸と佐渡おけさ
Sotokaifu Coast and Local Dancer

P201 弥彦山と越後平野
Mt. Yahiko and Echigo Plain

P200 相川 Aikawa
P201 弥彦 Yahiko

		**	●
P200	10 yen 多色 multicolored	100	40
P201	10 yen 多色 multicolored	80	40
P200-201 (2種 set of 2 values)		180	80
FDC (2種 set of 2 values)			600

切手データ Basic Information

発行枚数：各2,000万枚	number issued : 20 million each
製版・目打組合せ：タイプ1C-2	plate and perforation configuration : type 1C-2

P200 ●定常変種 Constant Flaws

A-pos.6　　A-pos.17　　A-pos.20　　B-pos.1

B-pos.13　　C-pos.2　　C-pos.13

D-pos.2

P201 ●定常変種 Constant Flaws

A-pos.10

B-pos.1　　B-pos.11

C-pos.5　　C-pos.16　　D-pos.5
　　C-pos.9　　　　　　　　D-pos.17

位置 position	特徴 feature	**	●
A-pos.6	マージンに黒点 black dot in bottom margin	300	100
A-pos.17	"便"の下方に青点 blue dot below "便"		
A-pos.20	小黒点と小青点 small black dot and small blue dot		
B-pos.1	帯の右に赤曲線 red arc to right of obi sash		
B-pos.13	岩の左に赤斜線 red line to left of rock		
C-pos.2	笠の左方に小青点 small blue dot to left of woven hat		
C-pos.13	赤点と小黒点 red dot and small black dot		
D-pos.2	黄線と極小点数個 yellow line and small dots		

位置 position	特徴 feature	**	●
A-pos.10	白ぼやけと右山に白点 white blur and white dot	200	100
B-pos.1	山に黄点 yellow dot on mountain		
B-pos.11	平野に青点 blue dot on plain		
C-pos.5	空に白ぼやけ white blur in sky		
C-pos.9	リタッチ風点群と緑点2ヵ所 retouched dots and two green dots		
C-pos.16	平野に小緑点 small green dot on plain		
D-pos.5	平野に小青点 small blue dot on plain		
D-pos.17	鳥のそばに黒点 black dot near birds		

国定公園切手

P202-203 秋吉台国定公園　発行日：1959（昭和34）.3.16
Akiyoshidai Quasi-National Park　Date of Issue：1959(Showa 34).3.16

P202 カルスト高原
Karst Plateau

P203 秋芳洞
Akiyoshi Cave

●風景印
Pictorial Postmarks

P202-203
秋吉 Akiyoshi

		**	●
P202	10 yen 多色 multicolored	120	40
P203	10 yen 多色 multicolored	240	40
	P202-203 (2種 set of 2 values)	360	80
	FDC (2種 set of 2 values)		650

切手データ Basic Information

発行枚数：各1,500万枚	number issued : 15 million each
製版・目打組合せ：P202=タイプ1C-1、P203=タイプ1C-3	plate and perforation configuration : P202=type 1C-1, P203=type 1C-3

P202 ●定常変種 Constant Flaws

A-pos.8　　A-pos.17　　B-pos.2　　B-pos.10

B-pos.20　　　　C-pos.5

C-pos.8　　D-pos.5　　D-pos.10　　D-pos.14

位置 position	特徴 feature	**	●
A-pos.8	"本"の下方に紫線 purple line below "本"	300	100
A-pos.17	石灰岩柱の右に青点 blue dot to right of limestone pillar		
B-pos.2	高原に黒点 black dot on plateau		
B-pos.10	"郵"の下方に青線 blue line below "郵"		
B-pos.20	高原に黒点 black dot on plateau		
C-pos.5	丸状の赤点群 round red dots		
C-pos.8	高原に筋状の紫線と小紫点 streaky purple line and small purple dot on plateau		
D-pos.5	空に白点 white dot in sky		
D-pos.10	リタッチ風白線 retouched white line		
D-pos.14	空に小青点 small blue dot in sky		

P203 ●定常変種 Constant Flaws

A-pos.2　　A-pos.11　　B-pos.5　　B-pos.8

B-pos.18　　C-pos.9　　C-pos.13　　D-pos.11

D-pos.14

位置 position	特徴 feature	**	●
A-pos.2	"1"に濃い青線 dark blue line at "1"	400	100
A-pos.11	"便"の下方に濃い点 dark dots below "便"		
B-pos.5	茶点2個と濃い青点 two brown dots and blue dot		
B-pos.8	"日"の下方に茶点 brown dot below "日"		
B-pos.18	"0"に茶点 brown dot at "0"		
C-pos.9	上端欠け nick in top edge		
C-pos.13	"吉"の下に小白点 small white dot below "吉"		
D-pos.11	少女の足元に白点 white dot near girl's feet		
D-pos.14	白点2ヵ所と濃い小点 two white dots and small dark dot		

国定公園切手

P204-205 耶馬日田英彦山国定公園
Yaba-Hita-Hikosan Quasi-National Park

発行日：1959（昭和34）.9.25
Date of Issue：1959(Showa 34).9.25

P204 カルスト高原 耶馬渓（青の洞門）Ao Cave Area of Yabakei

P205 日田・三隅川の鵜飼 Hita, Misumi River and Great Cormorant

●風景印 Pictorial Postmarks

P204 耶馬渓 Yabakei　　P205 日田 Hita

		**	●
P204	10 yen 多色 multicolored	140	60
P205	10 yen 多色 multicolored	200	60
P204-205	(2種 set of 2 values)	340	120
FDC	(2種 set of 2 values)		650

切手データ Basic Information	
発行枚数：各1,200万枚	number issued：12 million each
製版・目打組合せ：タイプ1C-2	plate and perforation configuration：type 1C-2

P204 ●定常変種 Constant Flaws

A-pos.2　A-pos.5　A-pos.13　B-pos.3
C-pos.1　C-pos.7　D-pos.12　D-pos.17

位置 position	特徴 feature	**	●
A-pos.2	山に紫点 purple dot on mountain	300	150
A-pos.5	"日"の上に青点 blue dot above "日"		
A-pos.13	"日"の右に青点 blue dot to right of "日"		
B-pos.3	"郵"の上方に小白点 small white dot above "郵"		
C-pos.1	"郵"の下に3連青点 three blue dots below "郵"		
C-pos.7	"便"の上方に薄い紫点 purple dot above "便"		
D-pos.12	"耶"の左手に濃い紫点 dark purple dot to left of "耶"		
D-pos.17	"耶"の左手に橙点 orange dot to left of "耶"		

P205 ●定常変種 Constant Flaws

A-pos.1　A-pos.20　B-pos.14　B-pos.17　C-pos.1

C-pos.6

D-pos.1　D-pos.18

位置 position	特徴 feature	**	●
A-pos.1	青斜線と小青点 blue line and small blue dots	400	150
A-pos.20	青斜線 diagonal blue line		
B-pos.14	"郵"の上に濃い青線 dark blue line above "郵"		
B-pos.17	空に青点 blue dot in sky		
C-pos.1	鵜近くに青点群 blue dots near cormorant		
C-pos.6	黒斜線とくちばし上に橙点 black line and orange dot on beak		
D-pos.1	空に青斜線 diagonal blue line in sky		
D-pos.18	"本"の上方に白点 white dot above "本"		

121

P206 三河湾国定公園　発行日：1960（昭和35）.3.20
Mikawawan (Bay) Quasi-National Park　Date of Issue：1960 (Showa 35) .3.20

●風景印
Pictorial Postmarks

蒲郡 Gamagori

P206 蒲郡海岸の竹島
Takeshima, off Gamagori

	**	●	FDC
P206 10 yen 多色 multicolored	120	60	400

切手データ Basic Information	
発行枚数：1,000万枚	number issued：10 million
製版・目打組合せ：タイプ1C-2	plate and perforation configuration：type 1C-2

P206 ●定常変種 Constant Flaws

 A-pos.4
 A-pos.14
 B-pos.7
 D-pos.18
 D-pos.18
 B-pos.13

B-pos.19　C-pos.6　D-pos.5

位置 position	特徴 feature	**	●
A-pos.4	青点2ヵ所 two blue dots	300	150
A-pos.14	海に青点 blue dot on sea		
B-pos.7	マージンに青点 blue dot in right margin		
B-pos.13	橋の上方に青点 blue dot above bridge		
B-pos.19	マージンに青点 blue dot in left margin		
C-pos.6	マージンに赤点 red dot in bottom margin		
D-pos.5	青点2ヵ所 two blue dots		
D-pos.18	丸く白抜け white spot		

P207 網走国定公園　発行日：1960（昭和35）.6.15
Abashiri Quasi-National Park　Date of Issue：1960 (Showa 35) .6.15

P207 濤沸湖畔の原生花園と斜里岳
Wild Flower Garden (Genseikaen) and Mt.Shari

●風景印
Pictorial Postmarks

網走 Abashiri

	**	●	FDC
P207 10 yen 多色 multicolored	200	80	450

切手データ Basic Information	
発行枚数：800万枚	number issued：8 million
製版・目打組合せ：タイプ1C-2	plate and perforation configuration：type 1C-2

P207 ●定常変種 Constant Flaws

 A-pos.8
 A-pos.17
 B-pos.2

B-pos.14

 C-pos.2
 C-pos.4
 D-pos.1
 D-pos.18

位置 position	特徴 feature	**	●
A-pos.8	青点3ヵ所 three blue dots	400	200
A-pos.17	"便"の下方に青点4個 four blue dots below "便"		
B-pos.2	マージンに黒線 black line in bottom margin		
B-pos.14	青い斜線、赤点、青点 blue line, red dot and blue dot		
C-pos.2	"日"の下に青斜線 blue line below "日"		
C-pos.4	マージン近くに黒点群 black dots near top margin		
D-pos.1	"便"の右に青点 blue dot to right of "便"		
D-pos.18	"日"の下に赤点群 red dots below "日"		

P208 足摺国定公園　発行日：1960（昭和35）.8.1
Ashizuri Quasi-National Park　Date of Issue：1960 (Showa 35) .8.1

P208 足摺岬灯台と巡礼の母娘
Lighthouse at Cape Ashizuri, Pilgrim Mother and Daughter

●風景印 Pictorial Postmarks

土佐清水
Tosashimizu

B-pos.17

D-pos.4　　D-pos.14

		**	●	FDC
P208 10 yen 多色 multicolored		120	80	450

	切手データ Basic Information	
発行枚数：800万枚	number issued：8 million	
製版・目打組合せ：タイプ1C-2	plate and perforation configuration：type 1C-2	

P208 ●定常変種 Constant Flaws

A-pos.1　　A-pos.3　　B-pos.1　　B-pos.5

位置 position	特徴 feature	**	●
A-pos.1	"日"の左に赤点 red dot to left of "日"	300	150
A-pos.3	海にリタッチ風濃い点 retouch-like dark dot on sea		
B-pos.1	マージンに小黒点 small black dot in SW margin		
B-pos.5	"郵"の下方に小紫点 small purple dot below "郵"		
B-pos.17	紫の横線2本 two purple horizontal lines		
D-pos.4	上腕に小紫点 small purple dot on upper arm		
D-pos.14	衣装に黒点 black dot on clothes		

P209 南房総国定公園　発行日：1961（昭和36）.3.15
Minami Boso Quasi-National Park　Date of Issue：1961 (Showa 36) .3.15

P209 白浜の野島崎灯台と海女
Nojimazaki Lighthouse and Women Divers

●風景印 Pictorial Postmarks

白浜 Shirahama

C-pos.20

D-pos.8

		**	●	FDC
P209 10 yen 多色 multicolored		80	60	400

	切手データ Basic Information	
発行枚数：800万枚	number issued：8 million	
製版・目打組合せ：タイプ1C-2	plate and perforation configuration：type 1C-2	

P209 ●定常変種 Constant Flaws

A-pos.16　　A-pos.20　　B-pos.5　　C-pos.2

位置 position	特徴 feature	**	●
A-pos.16	海に白ぼやけ white blur on sea	300	150
A-pos.20	マージンに黒点 black dot in bottom margin		
B-pos.5	マージンに青点 blue dot in left margin		
C-pos.2	"便"の右方に青点 blue dot to right of "便"		
C-pos.20	青線と小さな白点 blue line and small white dot		
D-pos.8	"本"の上に大きい黄点 large yellow dot above "本"		

P210 琵琶湖国定公園　Biwako (Lake) Quasi-National Park
発行日：1961（昭和36）.4.25　Date of Issue：1961 (Showa 36) .4.25

P210 石山寺観月亭からの琵琶湖と比叡の山々
Lake Biwa and Hiei Mountains from Ishiyama Temple

●風景印 Pictorial Postmarks

石山 Ishiyama

	**	●	FDC
P210 10 yen 多色 multicolored	80	60	400

切手データ Basic Information
発行枚数：800万枚	number issued：8 million
製版・目打組合せ：タイプ1C-2	plate and perforation configuration：type 1C-2

P210 ●定常変種 Constant Flaws

A-pos.1

A-pos.16

B-pos.5

C-pos.3

D-pos.13

位置 position	特徴 feature	**	●
A-pos.1	空に濃い点と白ぼやけ dark dot and white blur	200	150
A-pos.16	"便"の右方に青点 blue dot to right of "便"		
B-pos.5	右マージンに茶点2ヵ所（耳紙） two brown dots on NE selvage		
C-pos.3	"湖"の上に黒点 black dot above "湖"		
D-pos.13	"琵"と"琶"の間に濃い小点 dark small dot between "琵" and "琶"		

P211 山陰海岸国定公園　San'in Kaigan (Coast) Quasi-National Park
発行日：1961（昭和36）.8.15　Date of Issue：1961 (Showa 36) .8.15

P211 鳥取砂丘と因幡の傘踊り
Parasol Dance on Tottori Sand Dunes

●風景印 Pictorial Postmarks

鳥取 Tottori

	**	●	FDC
P211 10 yen 多色 multicolored	80	60	350

切手データ Basic Information
発行枚数：800万枚	number issued：8 million
製版・目打組合せ：タイプ2B-2	plate and perforation configuration：type 2B-2

1) 国定公園切手で初めての連続櫛型目打で、目打穿孔は二連2型が使用された。
1) This is the first Quasi-National Park Stamp perforated with the continuous comb perforator. The perforation was the double comb with two extension holes.

P211 ●定常変種 Constant Flaws

A-pos.6

A-pos.11

C-pos.16

D-pos.20

B-pos.11

B-pos.1

位置 position	特徴 feature	**	●
A-pos.6	"郵"の下方に小白点 small white dot below "郵"	200	150
A-pos.11	"郵"の右上に赤点 red dot to upper right of "郵"		
B-pos.1	腹部の左に赤点 red dot to left of stomach		
B-pos.11	"鳥"の右上に小黒点 small black dot to upper right of "鳥"		
C-pos.16	空に黒点 black dot in sky		
D-pos.20	"山"の下方に小赤点 small red dot below "山"		

P212 大沼国定公園　発行日：1961（昭和36）.9.15
Onuma Quasi-National Park　Date of Issue：1961 (Showa 36) .9.15

●風景印 Pictorial Postmarks

P212 大沼と駒ヶ岳
Onuma Lake and Komagatake Volcano

 大沼 Onuma

| | B-pos.5 | C-pos.4 | D-pos.1 | D-pos.18 |

		**	●	FDC
P212 10 yen 多色 multicolored		80	60	350

切手データ Basic Information	
発行枚数：800万枚	number issued：8 million
製版・目打合せ：タイプ1C-2	plate and perforation configuration：type 1C-2

位置 position	特徴 feature	**	●
A-pos.5	"便"の上方に小緑点 small green dot above "便"	200	150
A-pos.6	山の上方に小青点 small blue dot above mountain		
A-pos.14	白抜けと小青点 white spot and small blue dot		
B-pos.5	白抜けと緑の半円 white spot and green semicircle		
C-pos.4	"本"の上方に小青点 small blue dot above "本"		
D-pos.1	"便"の右下方に小緑点 small green dot to lower left of "便"		
D-pos.18	マージンに小緑点2個 two small green dots in bottom margin		

P212 ●定常変種 Constant Flaws

A-pos.5　A-pos.6　A-pos.14

P213 北長門海岸国定公園　発行日：1962（昭和37）.2.15
Kitanagato Kaigan (Coast) Quasi-National Park　Date of Issue：1962 (Showa 37) .2.15

P213 青海島
Omijima Island

●風景印 Pictorial Postmarks

 長門 Nagato

		**	●	FDC
P213 10 yen 多色 multicolored		80	60	300

切手データ Basic Information	
発行枚数：800万枚	number issued：8 million
製版・目打合せ：タイプ2B-2	plate and perforation configuration：type 2B-2

P213 ●定常変種 Constant Flaws

A-pos.9　A-pos.14　B-pos.9　B-pos.19　C-pos.2　C-pos.5　D-pos.4　D-pos.12

位置 position	特徴 feature	**	●
A-pos.9	青点4ヵ所 four blue dots	200	150
A-pos.14	海に小赤点 small red dot on sea		
B-pos.9	岩の上方に青点 blue dot above rock		
B-pos.19	小青点2ヵ所 two small blue dots		
C-pos.2	青点と小青点 blue dot and small blue dot		
C-pos.5	"郵"の下方に青点 blue dot below "郵"		
D-pos.4	"便"の下方に青曲線 blue arc below "便"		
D-pos.12	青点と小青点 blue dot and small blue dot		

1）この切手以降の国定公園切手はすべて連続櫛型目打による穿孔（タイプ2B-1、タイプ2B-2）となった。

1) This Quasi-National Park stamp and those that followed were perforated with the continuous comb perforator. The plate and perforation configuration of this stamp and thereafter was type 2B-1 or type 2B-2.

国定公園切手

P214 錦江湾国定公園　発行日：1962（昭和37）.4.30
Kinkowan (Bay) Quasi-National Park　Date of Issue：1962 (Showa 37) .4.30

P214 噴煙を上げる桜島と錦江湾
Sakurajima Volcano and Kinkowan Bay

●風景印 Pictorial Postmarks

鹿児島 Kagoshima

	**	●	FDC
P214 10 yen 多色 multicolored	60	60	300

切手データ Basic Information	
発行枚数：800万枚	number issued：8 million
製版・目打組合せ：タイプ2B-2	plate and perforation configuration：type 2B-2

P214 ●定常変種 Constant Flaws

A-pos.16　　B-pos.20　　D-pos.20

位置 position	特徴 feature	**	●
A-pos.16	右空に白点 white dot in sky on right	200	100
B-pos.20	"本"の右に小黒点 small black dot to right of "本"		
D-pos.20	噴煙に濃い線 dark line on plume		

P215 金剛生駒国定公園　発行日：1962（昭和37）.5.15
Kongo-Ikoma Quasi-National Park　Date of Issue：1962 (Showa 37) .5.15

P215 金剛山 Mt. Kongo

●風景印 Pictorial Postmarks

千早 Chijaya

	**	●	FDC
P215 10 yen 多色 multicolored	60	60	300

切手データ Basic Information	
発行枚数：800万枚	number issued：8 million
製版・目打組合せ：タイプ2B-2	plate and perforation configuration：type 2B-2

●目打穿孔 Perforation Method

図1 fig.1 二連2型 comb perforation with two extension holes
図2 fig.2 二連1型 comb perforation with one extension hole

図1 fig.1　図2 fig.2

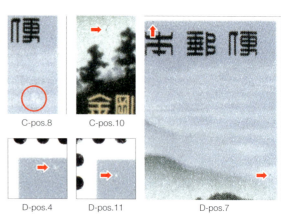

C-pos.8　C-pos.10
D-pos.4　D-pos.11　D-pos.7

P215 ●定常変種 Constant Flaws

A-pos.4　　A-pos.18　　B-pos.1

位置 position	特徴 feature	**	●
A-pos.4	"本"の下に小白点 small white dot below "本"	200	100
A-pos.18	マージンに緑点2ヵ所 two green dots in right and left margins		
B-pos.1	山に白点 white dot on mountain		
C-pos.8	"便"の下方に白抜け white spot below "便"		
C-pos.10	"金剛"の上方に2連点 two dots above "金剛"		
D-pos.4	第1コーナーに小白点 small white dot in NW corner		
D-pos.7	小黒点と灰点 small black dot and gray dot		
D-pos.11	第1コーナーに白点 white dot in NW corner		

1) この切手の目打穿孔は連続櫛型目打の二連2型と二連1型が併用されたが、それぞれの割合は不明である。

1) Two types of continuous comb perforator ware used for this stamp, the comb perforation with two extension holes and the comb perforation with one extension hole. The relative quantity of these two perforation types is unknown.

P216 水郷国定公園　Suigo Quasi-National Park

発行日：1962（昭和37）.6.1　Date of Issue：1962 (Showa 37) .6.1

P216 水郷景観と菖蒲 Suigo Park Scene and Iris

●風景印 Pictorial Postmarks

潮来 Itako

	**	●	FDC
P216 10 yen 多色 multicolored	60	60	350

切手データ Basic Information

発行枚数：800万枚	number issued：8 million
製版・目打組合せ：タイプ2B-2	plate and perforation configuration：type 2B-2

P216 ●定常変種 Constant Flaws

A-pos.3

 A-pos.14

 B-pos.1

 C-pos.7

 C-pos.11

C-pos.12　C-pos.15　C-pos.16

位置 position	特徴 feature	**	●
A-pos.3	青点3ヵ所 three blue dots	200	100
A-pos.14	山の右に白抜け white spot to right of mountain		
B-pos.1	船の上に青線 blue line above boat		
C-pos.7	"本"の上に青点 blue dot above "本"		
C-pos.11	菖蒲に紫点 purple dot on iris		
C-pos.12	菖蒲に青点 blue dot on iris		
C-pos.15	"国"の下に青点 blue dot under "国"		
C-pos.16	木の上に青い曲線 blue curve above tree		

P217 石鎚国定公園　Ishizuchi Quasi-National Park

発行日：1963（昭和38）.1.11　Date of Issue：1963 (Showa 38) .1.11

P217 石鎚山 Mt. Ishizuchi

●風景印 Pictorial Postmarks

石鎚 Ishizuchi

	**	●	FDC
P217 10 yen 多色 multicolored	60	50	350

切手データ Basic Information

発行枚数：1,000万枚	number issued：10 million
製版・目打組合せ：タイプ2B-2	plate and perforation configuration：type 2B-2

P217 ●定常変種 Constant Flaws

 A-pos.18

B-pos.13

 C-pos.13

 D-pos.9

位置 position	特徴 feature	**	●
A-pos.18	"日"の右上に青点 blue dot to upper right of "日"	200	100
B-pos.13	"本"の下方に赤点 red dot below "本"		
C-pos.13	雲に小茶点 small brown dot in cloud		
D-pos.9	"国"の右上欠け broken "国"		

1) この切手以降、国定公園切手の目打穿孔は二連1型になった。
1) The perforation of this Quasi-National Park stamp and those following was the double comb perforation with one extension hole.

国定公園切手

P218 玄海国定公園　Genkai Quasi-National Park
発行日：1963（昭和38）.3.15　Date of Issue：1963（Showa 38）.3.15

P218 芥屋の大門 Keya-no-Oto Rock

●風景印 Pictorial Postmarks

芥屋 Keya

	**	●	FDC
P218 10 yen 多色 multicolored	60	40	280

切手データ Basic Information	
発行枚数：1,000万枚	number issued：10 million
製版・目打組合せ：タイプ 2B-2	plate and perforation configuration: type 2B-2

P218 ●定常変種 Constant Flaws

A-pos.8　B-pos.9　C-pos.19　D-pos.5

位置 position	特徴 feature	**	●
A-pos.8	"郵"の右に青点 blue dot to right of "郵"	200	100
B-pos.9	"郵"と"便"の間に小青点 small blue dot between "郵" and "便"		
C-pos.19	"0"の右に茶点 brown dot to right of "0"		
D-pos.5	岩の上方に青点2ヵ所 two blue dots above rock		

●変則目打 Irregular Perforration

右耳紙 right selvage

1) この切手の目打穿孔は連続櫛型目打の二連1型である。目打の変種として一部のシートに櫛歯中段の横目打右側の目打針が1本少なく植えられているものが存在する。

1) The perforation of this stamp was the continuous double comb perforation with one extension hole, of which the right extension hole in the middle row was omitted.

図1 fig.1　目打穴数 number of extension hole 0-1-0-1-0
図2 fig.2　目打穴数 number of extension hole 1-0-1-0-1

図1 fig.1　図2 fig.2

P219 伊豆七島国定公園　Izushichitou (Islands) Quasi-National Park
発行日：1963（昭和38）.12.10　Date of Issue：1963（Showa 38）.12.10

P219 八丈島とフェニックス Hachijo Island and Phoenix Tree

●風景印 Pictorial Postmarks

八丈島 Hachijojima

	**	●	FDC
P219 10 yen 多色 multicolored	50	30	280

切手データ Basic Information	
発行枚数：1,500万枚	number issued：15 million
製版・目打組合せ：タイプ2B-2	plate and perforation configuration：type 2B-2

●変則目打 Irregular Perforration

左耳紙 left selvage

1) この切手の目打穿孔は連続櫛型目打の二連1型である。目打の変種として一部のシートに櫛歯端の横目打左側の目打針が1本多く植えられているものが存在する。

1) The perforation of this stamp was the continuous double comb perforation with one extension hole, with one extra extension hole on the left side in the top and bottom rows.

図1 fig.1　目打穴数 number of extension holes 1-2-1-2-1
図2 fig.2　目打穴数 number of extension holes 2-1-2-1-2

図1　fig.1　図2　fig.2

P219 ●定常変種 Constant Flaws

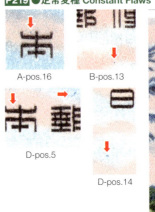

A-pos.16　B-pos.13

D-pos.5

D-pos.14

C-pos.16

位置 position	特徴 feature	**	●
A-pos.16	"本"の上に小青点 small blue dot above "本"	200	100
B-pos.13	"便"の下方に小青点 small blue dot below "便"		
C-pos.16	青曲線と山に大青点 blue arc and large blue dot		
D-pos.5	"本"に小青点と"郵"に青斜線 small blue dot and blue line		
D-pos.14	"日"の下に小青点 small blue dot below "日"		

P220 若狭湾国定公園　発行日：1964（昭和39）.1.25
Wakasawan (Bay) Quasi-National Park　Date of Issue : 1964 (Showa 39) .1.25

●風景印
Pictorial Postmarks

P220 高浜海岸と若葉山
Takahama Coast and Mt. Wakaba

高浜 Takahama

C-pos.18

C-pos.20

D-pos.9

D-pos.20

	**	●	FDC
P220 10 yen 多色 multicolored	50	30	280

切手データ Basic Information	
発行枚数：1,550万枚	number issued : 15.5 million
製版・目打組合せ：タイプ2B-2	plate and perforation configuration : type 2B-2

位置 position	特徴 feature	**	●
A-pos.8	"本"の下方に青点 blue dot below "本"	200	100
A-pos.9	マージンに青点 blue dot in top margin		
B-pos.2	右空に紫斜線 purple diagonal line in right sky		
B-pos.8	空に小紫点 small purple dot in sky		
C-pos.18	「つ」形状の青線 "つ" shaped curved blue line		
C-pos.20	「J」形状の青線 "J" shaped blue line		
D-pos.9	"便"の下方に青点線 blue dotted line below "便"		
D-pos.20	大きな紫点2個 two large purple dots		

P220 ●定常変種 Constant Flaws

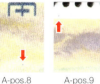
A-pos.8

A-pos.9

B-pos.2

B-pos.8

P221 日南海岸国定公園　発行日：1964（昭和39）.2.20
Nichinan Kaigan (Coast) Quasi-National Park　Date of Issue : 1964 (Showa 39) .2.20

●風景印
Pictorial Postmarks

P221 堀切峠からの波状岩（鬼の洗濯板）とリュウゼツラン
Wavy Rocks and Agave from Horikiri Pass

宮崎 Miyazaki

	**	●	FDC
P221 10 yen 多色 multicolored	50	30	280

切手データ Basic Information	
発行枚数：1,560万枚	number issued : 15.6 million
製版・目打組合せ：タイプ2B-2	plate and perforation configuration : type 2B-2

P221 ●定常変種 Constant Flaws

A-pos.8　A-pos.17　B-pos.14

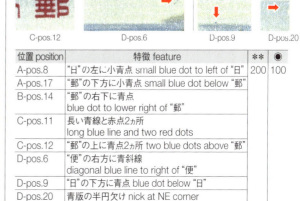

C-pos.12　D-pos.6　D-pos.9　D-pos.20

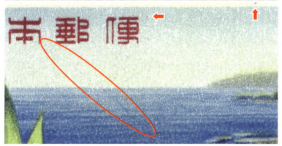
C-pos.11

位置 position	特徴 feature	**	●
A-pos.8	"日"の左に小青点 small blue dot to left of "日"	200	100
A-pos.17	"郵"の下方に小青点 small blue dot below "郵"		
B-pos.14	"郵"の右下に青点 blue dot to lower right of "郵"		
C-pos.11	長い青線と赤点2ヵ所 long blue line and two red dots		
C-pos.12	"郵"の上に青点2ヵ所 two blue dots above "郵"		
D-pos.6	"便"の右方に青斜線 diagonal blue line to right of "便"		
D-pos.9	"日"の下方に青点 blue dot below "日"		
D-pos.20	青版の半円欠け nick at NE corner		

P222 ニセコ積丹小樽海岸国定公園　発行日：1965（昭和40）.2.15
Niseko-Shakotan-Otaru Kaigan (Coast) Quasi-National Park　Date of Issue：1965 (Showa 40) .2.15

●風景印 Pictorial Postmarks

P222 ニセコアンヌプリ Mt. Niseko-Annupuri

昆布 Konbu

	**	●	FDC
P222 10 yen 多色 multicolored	50	30	280

切手データ Basic Information	
発行枚数：2,400万枚	number issued : 24 million
製版・目打組合せ：タイプ2B-2	plate and perforation configuration : type 2B-2

P222 ●定常変種 Constant Flaws

A-pos.1　A-pos.3　B-pos.14　B-pos.16

C-pos.2

位置 position	特徴 feature	**	●
A-pos.1	山頂上空に濃い青点 dark blue dot above mountaintop	200	100
A-pos.3	第1コーナーに白点 white dot in NW corner		
B-pos.14	"0"の中に茶点1個と黒点3個 brown dot and three black dots in "0"		
B-pos.16	"0"の中に茶点 brown dot in "0"		
C-pos.2	山間に濃い点 dark dot on mountain		

1) この切手以降、国定公園切手のスクリーン線数は250線となった。
1) After this stamp, 250-line screens were used for the Quasi-National Park stamps.

P223 蔵王国定公園　発行日：1966（昭和41）.3.15
Zao Quasi-National Park　Date of Issue：1966 (Showa 41) .3.15

●風景印 Pictorial Postmarks

P223 蔵王の火口湖 Crater Lake, Zao

遠刈田 Togatta

	**	●	FDC
P223 10 yen 多色 multicolored	50	30	280

切手データ Basic Information	
発行枚数：2,400万枚	number issued : 24 million
製版・目打組合せ：タイプ2B-2	plate and perforation configuration : type 2B-2

D-pos.7

P223 ●定常変種 Constant Flaws

A-pos.5　A-pos.14　B-pos.1　B-pos.5

C-pos.6　C-pos.7　D-pos.6

位置 position	特徴 feature	**	●
A-pos.5	マージンに赤点 red dot in right margin	200	100
A-pos.14	火口左に小青点 small blue dot to left of crater		
B-pos.1	マージンに茶点 brown dot in left margin		
B-pos.5	青点2ヵ所と「ノ」形状の青線 two blue dots and blue line		
C-pos.6	茶点群3ヵ所 three brown dots		
C-pos.7	マージンに青点2個 two blue dots in top margin		
D-pos.6	火口左に小赤点 small red dot to left of crater		
D-pos.7	青点2ヵ所と色抜け two blue dots and discolored spot		

1) この切手以降、国定公園切手のスクリーン角度が全色45度となった。
1) After this stamp, the screen angle of all colors for the Quasi-National Park stamps became 45°.

P224-225 室戸阿南海岸国定公園
Muroto-Anan Kaigan (Coast) Quasi-National Park

発行日：1966（昭和41）.3.22
Date of Issue : 1966 (Showa 41) .3.22

P224 室戸海岸 Muroto Cape

P225 阿南海岸の千羽海崖
Senba Cliffs, Anan Coast

● 風景印 Pictorial Postmarks

P224 室戸
Muroto

P225 日和佐
Hiwasa

	**	●
P224 10 yen 多色 multicolored	50	30
P225 10 yen 多色 multicolored	50	30
P224-225 (2種 set of 2 values)	100	60
FDC (2種 set of 2 values)		560

切手データ Basic Information	
発行枚数：各2,150万枚	number issued : 21.5 million each
製版・目打組合せ：P224=タイプ2B-2、P225=タイプ2B-1	plate and perforation configuration : P224=type 2B-2, P225=type 2B-1

P224 ● 定常変種 Constant Flaws

A-pos.1

A-pos.11

B-pos.4

C-pos.6

D-pos.18

位置 position	特徴 feature	**	●
A-pos.1	左の山の上空に青点群 blue dots in sky above mountain on left	200	100
A-pos.11	"便"の下方に黒点2ヵ所 two black dots below "便"		
B-pos.4	"阿"の上に青点 blue dot above "阿"		
C-pos.6	海に濃い青点 dark blue dot on sea		
D-pos.18	マージンに小青点 small blue dot in left margin		

P225 ● 定常変種 Constant Flaws

A-pos.17

A-pos.20

B-pos.1

B-pos.9

B-pos.4

C-pos.1

C-pos.8

D-pos.2

D-pos.7

位置 position	特徴 feature	**	●
A-pos.17	空に黒点2ヵ所 two black dots in sky	200	100
A-pos.20	雲に黒点 black dot on cloud		
B-pos.1	"0"の上に白抜け white spot above "0"		
B-pos.4	空に黒点 black dot in sky		
B-pos.9	マージンに破線 broken line in bottom margin		
C-pos.1	"N"に大きな白点、第2マージンに黒点 large white dot on "N", black dot in NE margin		
C-pos.8	"10"の右下に赤点 red dot to lower right of "10"		
D-pos.2	マージンに黒点 black dot in bottom margin		
D-pos.7	崖の左に黒点 black dot to left of cliff		

年賀切手

年賀切手 New Year's Greeting Stamps 1952-1966

■発行の経緯

年賀郵便の取扱通数が増加する中、私製はがきの年賀状に貼る専用切手として「年賀切手」が発行された。また「お年玉くじつき年賀はがき」の賞品に充てるため、「お年玉小型シート」が1950年(昭和25)から発行された。

本カタログでは、1952年(昭和27)から1965年(昭和40)に発行された年賀切手(N7〜N21)とお年玉小型シート(N7A〜N21A)を採録している。

■製版・目打組合せ、刷色とスクリーン角度

年賀切手はN7から単片20枚シートとなり、印刷機はN7〜N9が板グラビア機、N10以降はグラビア輪転機が使用された。実用版構成はN7〜N13が2×2面頭合せ、N14〜N17が2×2面順並び、N18以降は4面横並びである。目打穿孔はN7〜N14が2面横並び(タイプ1A-4、1B-4)、N15〜N17が4面(タイプ1C-4)、N18以降は横連続(タイプ2B-1)である。

小型シートはすべて単片4枚構成に変更され、印刷機は年賀切手と同様である。実用版構成はN7A〜N9Aが3×4の12面、N10A以降は2×6の12面である。目打はN17Aまでは全型目打、または全型目打と櫛型目打で上段・下段の2度穿孔された。そのため、中央の横目打が不整になったものがある。N18A以降は、田型状の全型目打によって連続的に一度に穿孔するようになった。

20枚シートと小型シートの目打はどちらもピッチ13½であるため、単片の場合は耳紙つきでないと識別が難しい。ただし20枚シートが横紙であるのに対してN7A、N18A〜N21Aは縦紙なので、用紙の反りで識別が可能である。

なお、現在判明している刷色とスクリーン角度の組合せはp.143に示した。これにより実用版の分類、20枚シートと小型シートの識別が可能となったが、測定誤差も考えられるため注意が必要である。小型シートは複数の実用版が存在して定常変種も多く研究途上のため、定常変種の掲載は見送った。

●小型シートの目打穿孔 (N7A〜N17A)
Perforation Method of Souvenir Sheet (N7A〜N17A)

上段 top stamps / 下段 bottom stamps

全型目打 harrow perf.　　全型＋櫛型目打 harrow and comb perf.

■ Issuance of New Year's Greeting Stamps

As the volume of New Year Greeting mail increased, "New Year's Greeting Stamps" were issued as special stamps for use on New Year's Greeting postcards. Additionally, "New Year's Souvenir Sheets" were issued from 1950 (Showa 25) to be used as prizes for the "New Year's Greeting Lottery Postal Cards" with a lottery number.

This catalogue includes New Year's Greeting stamps (N7 - N21) and New Year's Souvenir Sheets (N7A - N21A) issued from 1950 (Showa 27) to 1965 (Showa 40).

■ Plate and Perforation Configurations, Color Shades and Screen Angles

Beginning with N7, the number of stamps in a pane for the New Year's Greeting stamps became twenty. The flat plate photogravure press was used for N7 - N9, and the rotary photogravure press was used for N10 and thereafter. Pane configurations of plate were as follows; 2×2-pane head-to-head configuration for N7 - N13, and 2×2-pane normal configuration for N14 - N17, 4-pane vertical configuration for N18 and thereafter. Pane configurations for perforation were as follows: 4-pane horizontal configuration (types 1A-4 and 1B4) for N7 - N14, 2×2-pane configuration (type 1C-4) for N15 – N17, and horizontal configuration web paper (type 2B-1) for N18 and thereafter.

The stamp configuration in the souvenir sheet was changed to 2×2, and the press was same as for the sheet New Year's Greeting stamps. The pane configurations of the plate were as follows: 3×4 for N7A – N9A, and 2×6 for N10A and thereafter. Harrow perforation, or a combination of harrow and comb perforation, was used for N7A - N17A. In instances of a combination perforation, shifted perforations are known for some souvenir sheets. Only the harrow perforation was used for N18A and thereafter.

As the perforation gauge for both the sheet stamps and souvenir sheet stamps was 13½, it is difficult to determine whether a single stamp without selvage comes from a full sheet or a souvenir sheet. However, since the sheet stamps were printed on short grain paper while N7A and N18A - N21A were printed on long grain paper, it can be determined whether these stamps come from a full sheet or souvenir sheet by paper grain.

The combinations of color shades and screen angles are shown in the table on p. 143. By using this table, it is possible to classify plates and distinguish sheet stamps from souvenir sheet stamps. However, since measurement of screen angle is subject to error, care must be taken in measurement. As more than one plate was used for the souvenir sheets, and there were many constant flaws, the plate identifications of the souvenir sheets remains under investigation.

共通切手データ Common Information

N7 - N21	
版式: N7-N9＝板グラビア版、N10-N21＝グラビア輪転版	printing plate: N7-N9=flat photogravure plate, N10-N21=rotary photogravure plate
目打: 櫛型13½	perforation: comb perf. 13½
印面寸法: 22.5×27mm	printing area size: 22.5×27mm
シート構成: 20枚(横4×縦5)	pane format: 20 subjects (4×5)
銘版、銘版位置:N7=⑫新庁銘19番、N8-N21=⑬大蔵銘19番	imprint and position: N7=type ⑫ pos.19, N8-N21=type ⑬ pos.19

N7A - N21A	
版式: N7A-N9A＝板グラビア版、N10A-N21A＝グラビア輪転版	printing plate: N7A-N9A=flat photogravure plate, N10A-N21A=rotary photogravure plate
目打: N7A-N17A＝全型13½、全型櫛型13½、N18A-N21A＝全型13½	perforation : N7A-N17A=harrow perf. 13½, harrow and comb perf. 13½, N18A-N21A= harrow perf. 13½
シート寸法: N7A-N17A=102×90mm、N18A-N21A=102×93.5mm	sheet size: N7A-N17A=102×90mm, N18A-N21A=102×93.5mm

年賀切手

N7/N7A 1952年（昭和27）用
For 1952 (Showa 27)

発行日：N7＝1952（昭和27）.1.16、N7A＝1952（昭和27）.1.20
Date of Issue：N7=1952 (Showa 27).1.16, N7A=1952 (Showa 27).1.20

N7 翁の能面
Noh Mask of "Okina"

N7A お年玉小型シート
"New Year's Gift" Souvenir Sheet (lottery prize)

	**	●	FDC
N7 5 yen 紅 crimson rose	2,500	200	6,500
N7A 20 yen (5 yen × 4)	27,000	27,000	28,000

切手データ Basic Information

N7
発行枚数：1,000万枚　number issued : 10 million
製版・目打組合せ：タイプ1A-4　plate and perforation configuration : type 1A-4

N7A
発行枚数：308.7万枚　number issued : 3,087,000

N7 ●定常変種 Constant Flaws

A-pos.6 | A-pos.15/19 | C-pos.16 | A-pos.18
C-pos.20 | D-pos.19

位置 position	特徴 feature	**	●
A-pos.6	"5"の左下に濃い点 dark dot to lower left of "5"	6,000	500
A-pos.15/19	pos.15はバーの右下に白抜け、pos.19は第2コーナーに大きな濃い点 white spot to lower right of bar at pos.15, large dark dot in NE corner at pos.29		
A-pos.18	"日"の上に小さな点、能面の左に大きな濃い点、髭の先に小さな点 small dot above "日", large dark dot to left of Noh Mask, small dot at tip of beard		
C-pos.16	"便"の右上に大きな濃い点 large dark dot to upper right of "便"		
C-pos.20	第4コーナーに濃い点、"5"の下に点 dark dot in SW corner, dot below "5"		
D-pos.19	"2"の右に濃い点、"5"に小さな点2個 dark dot to right of "2", small two dots near "5"		

N8/N8A 1953年（昭和28）用
For 1953 (Showa 28)

発行日：N8＝1953（昭和28）.1.1、N8A＝1953（昭和28）.1.20
Date of Issue：N8=1953 (Showa 28).1.1, N8A=1953 (Showa 28).1.20

N8 御所人形「三番叟」
Court Doll of "Sanbasou"

N8A お年玉小型シート
"New Year's Gift" Souvenir Sheet (lottery prize)

●年賀印
New Year Postmark
彦根 HIkone

	**	●	FDC
N8 5 yen 赤 carmine	1,700	200	5,500
N8A 20 yen (5 yen × 4)	18,000	18,000	19,000

切手データ Basic Information

N8
発行枚数：500万枚　number issued : 5 million
製版・目打組合せ：タイプ1A-4　plate and perforation configuration : type 1A-4

N8A
発行枚数：315万枚　number issued : 3,150,000

1) N8Aは、左側タブ付である。
1) Tabs are attached to the left of the stamps in N8A.

N8 ●定常変種 Constant Flaws

A-pos.7

 A-pos.11
 B-pos.5
 B-pos.18
 C-pos.1

 C-pos.18
 D-pos.8
 D-pos.19
 D-pos.20

年賀切手

位置 position	特徴 feature	**	●
A-pos.7	"日"の左下に大きな白点、"1953"の周囲に点多数 large white dot to lower left of "日", many dots around "1953"	5,000	500
A-pos.11	"5"の左下に濃い横線 dark horizontal line to lower left of "5"		
B-pos.5	腹に大きな赤い点 large red dot on stomach		
B-pos.18	"5"の左下に帯状の縦線 vertical line to lower left of "5"		
C-pos.1	足に濃い太い線 thick dark lines on leg		
C-pos.18	小指に濃い点 dark dot on little finger		
D-pos.8	足に濃い太い線 thick dark lines on leg		
D-pos.19	膝に濃い点、腹に点 dark dot on knee, dot on stomach		
D-pos.20	"5"の左に濃い3連点、5の右に帯線 three dark dots to left of "5", band line to right of "5"		

N9/N9A 1954年（昭和29）用 For 1954 (Showa 29)

発行日：N9=1953（昭和28）.12.25、N9A=1954（昭和29）.1.20
Date of Issue：N9=1953 (Showa 28).12.25, N9A=1954 (Showa 29).1.20

N9 木工品「三春駒」（福島の郷土玩具）
Wooden Toy "Miharu-koma" (Folk Toy of Fukushima)

N9A お年玉小型シート
"New Year's Gift" Souvenir Sheet (lottery prize)

切手データ Basic Information

N9	
発行枚数：500万枚	number issued：5 million
製版・目打組合せ：タイプ1A-4	plate and perforation configuration：type 1A-4
N9A	
発行枚数：363.1万枚	number issued：3,631,000

N9 ●定常変種 Constant Flaws

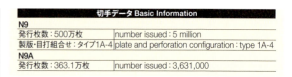

B-pos.1　　　　　　C-pos.17

位置 position	特徴 feature	**	●
B-pos.1	"便"の右下方に赤点 red dot to lower right of "便"	3,000	400
C-pos.17	"昭和二十"の右に小さな赤点 small dot to right of "昭和二十"		

●風景印 Picture Postmark

三春 Miharu

	**	●	FDC
N9 5 yen 赤 rose	1,200	150	3,500
N9A 20 yen (5 yen × 4)	10,000	10,000	11,000

N10/N10A 1955年（昭和30）用 For 1955 (Showa 30)

発行日：N10=1954（昭和29）.12.20、N10A=1955（昭和30）.1.20
Date of Issue：N10=1954 (Showa 29).12.20, N10A=1955 (Showa 30).1.20

N10 加賀八幡起き上がり（金沢の郷土玩具）
Kaga Hachiman Okiagari Doll (Folk Toy of Kanazawa)

N10A お年玉小型シート
"New Year's Gift" Souvenir Sheet (lottery prize)

切手データ Basic Information

N10	
発行枚数：500万枚	number issued：5 million
製版・目打組合せ：タイプ1A-4	plate and perforation configuration：type 1A-4
N10A	
発行枚数：400万枚	number issued：4 million

1) 2色刷グラビア輪転版による記念特殊切手は、N10が最初である。スクリーン線数は260線となり、スクリーン角度は刷色ごとに異なる値になった。
1) N10 is the first two-color stamp printed on the two-color photogravure rotary press. The screen number is 260 and the screen angle is different for each color.

●風景印 Picture Postmark

金沢 Kanazawa

	**	●	FDC
N10 5 yen 黄味赤・灰黒 red and black	1,200	150	4,000
N10A 20 yen (5 yen × 4)	10,000	10,000	12,000

N10 ●定常変種 Constant Flaws

A-pos.20

B-pos.2

B-pos.6

C-pos.2

C-pos.6

D-pos.4

D-pos.18

位置 position	特徴 feature	**	●
A-pos.20	第4コーナーに黒点 black dot in SW corner	3,000	400
B-pos.2	第4コーナーに黒点3個 three black dots in SW corner		
B-pos.6	眼の下に黒点 black dot below eye		
C-pos.2	右マージンに赤縦線 vertical red line in right margin		
C-pos.6	第4コーナーに黒点4個 four black dots in SW corner		
D-pos.4	眼の下に黒点 black dot below eye		
D-pos.18	"便"の上方に黒点、下方に赤点 black dot above and red dot below "便"		

N11/N11A　1956年（昭和31）用 / For 1956 (Showa 31)

発行日：N11=1955（昭和30）.12.20、N11A=1956（昭和31）.1.20
Date of Issue：N11=1955 (Showa 30) .12.20, N11A=1956 (Showa 31) .1.20

N11 こけし
（東北地方の郷土玩具）
Kokeshi Dolls
(Folk Toy of
Tohoku Region)

N11A お年玉小型シート
"New Year's Gift" Souvenir
Sheet (lottery prize)

●年賀印
Picture
Postmark

吉川 Yoshikawa

		**	●	FDC
N11 5 yen オリーブ緑・赤味だいだい olive green and red		500	70	3,000
N11A 20 yen (5 yen × 4)		7,000	7,000	8,500

切手データ Basic Information	
N11	
発行枚数：600万枚	number issued：6 million
製版・目打組合せ：タイプ1A-4	plate and perforation configuration：type 1A-4
N11A	
発行枚数：585.2万枚	number issued：5,852,000

1) 赤色の定常変種が確認できるものを1版、赤色の定常変種が確認できないものを2版に分類した。各刷色では、1版と2版に共通の定常変種は確認できていない。

1) The plate with constant flaws in the red plate is classified as plate 1, and the plate without constant flaws in the red plate is classified as plate 2.

N11 ● 定常変種 Constant Flaws

P1-A-pos.1

P1-A-pos.4

P1-B-pos.1

P1-C-pos.9

P1-B-pos.5/6

P1-C-pos.14

P1-D-pos.14

P1-D-pos.1

P2-A-pos.1

P2-B-pos.2

P2-B-pos.9

位置 position	特徴 feature	**	●
P1-A-pos.1	左耳紙から第1コーナーに赤線 red line from left selvage to NW corner of stamp	1,000	200
P1-A-pos.4	こけしに緑の斜線 green diagonal line on Kokeshi Doll		
P1-B-pos.1	左マージンから第4コーナーに赤い斜線 diagonal red line from left margin to SW corner		
P1-B-pos.5/6	pos.5からpos.6の上マージンに緑の斜線 diagonal green line in top margin of pos.5 and pos.6		
P1-C-pos.9	肩に赤点、肩の右に赤点 red dot on shoulder, red dot to right of shoulder		
P1-C-pos.14	顔に大きな緑点 large green dot on face		
P1-D-pos.1	耳紙と左マージンに緑線 two green lines in left selvage and left margin		
P1-D-pos.14	右マージンに赤点 red dot in right margin		
P2-A-pos.1	第1コーナーに緑地ベタ green space in NW corner		
P2-B-pos.2	第1コーナーに緑点 green dot in NW margin		
P2-B-pos.9	竹の葉に濃い緑点 dark green dot on bamboo leaf		

年賀切手

N12/N12A 1957年（昭和32）用
発行日：N12=1956（昭和31）.12.20、N12A=1957（昭和32）.1.20

For 1957 (Showa 32)　Date of Issue: N12=1956 (Showa 31).12.20, N12A=1957 (Showa 32).1.20

N12 鯨のだんじり
（長崎の郷土玩具）
Whale Toy
"Danjiri" (Folk
Toy of Nagasaki)

N12A お年玉小型シート
"New Year's Gift" Souvenir
Sheet (lottery prize)

B-pos.2　B-pos.6　C-pos.11　C-pos.16

C-pos.18　D-pos.1　D-pos.4

D-pos.10

● 風景印
Picture
Postmark
 長崎 Nagasaki

	**	●	FDC
N12 5 yen 多色 multicolored	300	50	2,300
N12A 20 yen (5 yen × 4)	3,000	3,500	6,000

切手データ Basic Information	
N12	
発行枚数：600万枚	number issued : 6 million
製版・目打組合せ：タイプ1A-4	plate and perforation configuration : type 1A-4
N12A	
発行枚数：860.6万枚	number issued : 8,606,000

N12 ● 定常変種 Constant Flaws

A-pos.1　　　　A-pos.5

位置 position	特徴 feature	**	●
A-pos.1	"本"の下に白点、鯨の上に白点2個 white dot below "本", two white dots above whale	1,000	200
A-pos.5	"本"の右下方に白点 white dot to lower right of "本"		
B-pos.2	左マージンに青点 blue dot in left margin		
B-pos.6	"5"の左に白抜け white spot to left of "5"		
C-pos.11	"本"の下に白点 white dot below "本"		
C-pos.16	車輪に青点2個 two blue dots on wheel		
C-pos.18	鯨の頭欠け nick on whale head		
D-pos.1	"便"の下方に大きな白点 large white dot below "便"		
D-pos.4	車輪に青線 blue line on wheel		
D-pos.10	車輪に大きな青点 large blue dot on wheel		

N13/N13A 1958年（昭和33）用
発行日：N13=1957（昭和32）.12.20、N13A=1958（昭和33）.1.20

For 1958 (Showa 33)　Date of Issue: N13=1957 (Showa 32).12.20, N13A=1958 (Showa 33).1.20

N13 犬張り子
（東京の郷土玩具）
Papier Mache Dog
(Folk Toy of Tokyo)

N13A お年玉小型シート
"New Year's Gift" Souvenir
Sheet (lottery prize)

	**	●	FDC
N13 5 yen 多色 multicolored	60	50	1,000
N13A 20 yen (5 yen × 4)	1,300	1,500	4,500

切手データ Basic Information	
N13	
発行枚数：800万枚	number issued : 8 million
製版・目打組合せ：タイプ1A-4	plate and perforation configuration : type 1A-4
N13A	
発行枚数：948.4万枚	number issued : 9,484,000

● 風景印
Picture
Postmark
 浅草 Asakusa

1) 赤色と黒色の定常変種が確認できるものを1版、赤色と黒色の定常変種が確認できないものを2版に分類した。各刷色では、1版と2版に共通の定常変種は確認できていない。

1) The plate with constant flaws in the red and black plates is classified as plate 1, and the plate without constant flaws in the red and black plates is classified as plate 2. No common constant flaws in the color plates have been identified between plates 1 and 2.

年賀切手

N13 ● 定常変種 Constant Flaws

P1-A-pos.16

P1-A-pos.17

P2-A-pos.4

P1-B-pos.1　　P1-B-pos.16

P2-A-pos.13

P2-A-pos.20

P1-C-pos.8　　P1-C-pos.18

P2-B-pos.5

P2-B-pos.6

P1-C-pos.9

P1-D-pos.10

位置 position	特徴 feature	**	●
P1-A-pos.16	髭の下に赤点 red dot below whiskers	200	100
P1-A-pos.17	第1コーナーのマージンに大きな黒点 large black dot in margin of NW corner		
P1-B-pos.1	印面左縁に濃い黒点と玉状の黒点 dark black dot and round black dot on left edge of printed area		
P1-B-pos.16	脚に赤点 red dot on leg		
P1-C-pos.8	胴に赤点 red dot on body		
P1-C-pos.9	"8"の左下に赤い汚れ、第3マージンに赤い斜線 red stain to lower left of "8", red diagonal line in SE margin		
P1-C-pos.18	"便"の右に濃い黒点 dark black dot to left of "便"		
P1-D-pos.10	印面左縁に黒点と黒点群 black dot and black dots on left edge of printed area		
P2-A-pos.4	印面左縁に黒点3個 three black dots on left edge of printing area		
P2-A-pos.13	頬に小さな緑点 small green dot on cheek		
P2-A-pos.20	頬に赤の横線、髭の下に赤点 red horizontal line on cheek, red dot below whiskers		
P2-B-pos.5	耳に緑点 green dot on ear		
P2-B-pos.6	"5"の中に白抜きとリタッチ white spot and retouching in "5"		

N14 / N14A　1959年（昭和34）用　For 1959 (Showa 34)

発行日：N14=1958（昭和33）.12.20、N14A=1959（昭和34）.1.20
Date of Issue：N14=1958 (Showa 33).12.20, N14A=1959 (Showa 34).1.20

N14 鯛えびす（高松の郷土玩具）
Papier Mache "Tai-ebishu" (Folk Toy of Takamatsu)

N14A お年玉小型シート
"New Year's Gift" Souvenir Sheet (lottery prize)

● 風景印 Picture Postmark　高松 Takamatsu

	**	●	FDC
N14　5 yen 多色 multicolored	60	40	800
N14A 20 yen (5 yen × 4)	1,300	1,500	3,500

切手データ Basic Information	
N14	
発行枚数：1,500万枚	number issued：15 million
製版・目打組合せ：タイプ1B-4	plate and perforation configuration：type 1B-4
N14A	
発行枚数：1,044.2万枚	number issued：10,442,000

1) 各刷色では、1版と2版に共通の定常変種は確認できていない。
1) No common constant flaw of each color between plates 1 and 2 is known.

● スクリーン角度 Screen Angles

1版 Plate 1
赤60度
red 60 degrees

2版 plate 2
赤65度
red 65 degrees

N14 ● 定常変種 Constant Flaws

P1-A-pos.1

P1-A-pos.12

P1-B-pos.11

P1-B-pos.19

P2-A-pos.1　　P2-A-pos.19

P2-B-pos.14　　P2-B-pos.16

年賀切手

P2-C-pos.6

P2-C-pos.13

P2-D-pos.11

P2-D-pos.6

位置 position	特徴 feature	**	●
P1-A-pos.1	目の下に黒点 black dot below eye	200	100
P1-A-pos.12	"5"の右方に青点 blue dot to right of "5"		
P1-B-pos.11	"5"の下に大きな白点 large white dot below "5"		
P1-B-pos.19	"5"の角欠け nick in "5"		
P2-A-pos.1	"便"の上方に白点 white dot above "便"		
P2-A-pos.19	"本"の下に黄茶点 yellow-brown dot below "本"		
P2-B-pos.14	第2コーナーのマージンに赤線 red line in margin of NE corner		
P2-B-pos.16	鯛えびすの右に赤点 red dot to right of "Tai-ebishu"		
P2-C-pos.6	唇の下に赤点 red dot below lip		
P2-C-pos.13	"日"の下方に赤点 red dot below "日"		
P2-D-pos.6	首に大きな青玉、左マージンに赤点 large blue spot on neck, red dot in left margin		
P2-D-pos.11	尾の上に赤点2個 two red dots above tail		

N15 / N15A　1960年（昭和35）用　For 1960 (Showa 35)

発行日：N15=1959（昭和34）.12.19、N15A=1960（昭和35）.1.20
Date of Issue：N15=1959 (Showa 34).12.19, N15A=1960 (Showa 35).1.20

N15 米食いねずみ
（金沢の郷土玩具）
Papier Mache Mouse
(Folk Toy of Kanazawa)

N15A お年玉小型シート
"New Year's Gift" Souvenir
Sheet (lottery prize)

●風景印
Picture Postmark

金沢
Kanazawa

	**	●	FDC
N15 5 yen 多色 multicolored	80	40	800
N15A 20 yen (5 yen × 4)	1,300	1,500	2,300

切手データ Basic Information	
N15	
発行枚数：1,000万枚	number issued：10 million
製版・目打組合せ：タイプ1C-4	plate and perforation configuration：type 1C-4
N15A	
発行枚数：1,101.7万枚	number issued：11,017,000

●スクリーン角度 Screen Angles

N15 金79度 gold 79 degrees

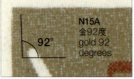
N15A 金92度 gold 92 degrees

1）金色のスクリーンは、網目が見えなくても印面周辺の段差により識別できる。段差の周期が角度90度前後の小型シートでは2.7mm以上あるのに対し、20枚シート切手では約0.5mmでギザギザのように見える。

1) Although the screen pattern cannot be observed, the screen angle of the gold plate can be identified from the wavy edge, . The "wave length" of the edge of the 90° screen is longer than 2.7 mm, wheras that of the 79°screen is about 0.5 mm.

N15 ●定常変種 Constant Flaws

A-pos.11

A-pos.15

B-pos.8

B-pos.19

C-pos.14

C-pos.15

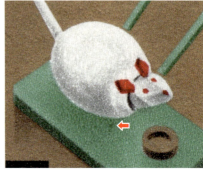
D-pos.18

位置 position	特徴 feature	**	●
A-pos.11	ねずみに黒点2個 two black dots on mouse	200	100
A-pos.15	ねずみに黒点 black dot on mouse		
B-pos.8	緑の板に濃い緑点 dark green dot on green board		
B-pos.19	"便"の字が切れ defect of "便"		
C-pos.14	緑の台の右に白抜け white spot to right of green board		
C-pos.15	緑の台に白点 white dot on green board		
D-pos.18	緑の台に染み stain on green board		

N16 / N16A 1961年（昭和36）用
発行日：N16=1960（昭和35）.12.20、N16A=1961（昭和36）.1.20
For 1961 (Showa 36)　　Date of Issue：N16=1960 (Showa 35).12.20, N16A=1961 (Showa 36).1.20

N16 赤べこ（会津地方）と金のべこっこ（岩手）
Akabeko Toy (Aizu Region in Fukushima) and Paper Mache Gold Calf (Iwate)

N16A お年玉小型シート
"New Year's Gift" Souvenir Sheet (lottery prize)

N16 ●定常変種 Constant Flaws

A-pos.6　　B-pos.8　　B-pos.10　　C-pos.9

C-pos.20　　D-pos.1　　D-pos.20

位置 position	特徴 feature	**	●
A-pos.6	足の下に小さな赤点3個 three red dots below foot	200	100
B-pos.8	赤べこの上に赤点 red dot above Akabeko		
B-pos.10	胴体に茶色抜け white spot on body		
C-pos.9	紐の下に赤点 red dot below string		
C-pos.20	"5"の下に白ぼやけ white blur below "5"		
D-pos.1	白色の台に緑点 green dot on white board		
D-pos.20	黄色の台の右下に白点 white dot to lower right of yellow board		

●風景印 Picture Postmark　会津若松 Aizuwakamatsu　盛岡 Morioka

	**	●	FDC
N16 5 yen 多色 multicolored	80	40	700
N16A 20 yen (5 yen × 4)	1,500	1,600	2,300

切手データ Basic Information	
N16	
発行枚数：800万枚	number issued：8 million
製版・目打組合せ：タイプ1C-4	plate and perforation configuration：type 1C-4
N16A	
発行枚数：1,079.6万枚	number issued：10,796,000

年賀

N17 / N17A 1962年（昭和37）用
発行日：N17=1961（昭和36）.12.15、N17A=1962（昭和37）.1.20
For 1962 (Showa 37)　　Date of Issue：N17=1961 (Showa 36).12.15, N17A=1962 (Showa 37).1.20

N17 張り子の虎（島根の郷土玩具）
Papier Mache Tiger (Folk Toy of Shimane)

N17A お年玉小型シート
"New Year's Gift" Souvenir Sheet (lottery prize)

切手データ Basic Information	
N17	
発行枚数：1,000万枚	number issued：10 million
製版・目打組合せ：タイプ1C-4	plate and perforation configuration：type 1C-4
N17A	
発行枚数：1,091.8万枚	number issued：10,918,000

N17 ●定常変種 Constant Flaws

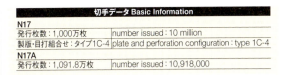

A-pos.2　　A-pos.11　　B-pos.17　　B-pos.18

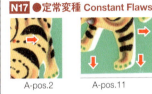

C-pos.1　　C-pos.10　　D-pos.7　　D-pos.13

●風景印 Picture Postmark　出雲 Izumo

	**	●	FDC
N17 5 yen 多色 multicolored	50	40	400
N17A 20 yen (5 yen × 4)	1,500	1,600	2,300

年賀切手

位置 position	特徴 feature	**	●
A-pos.2	胴体に赤点 red dot on body	200	100
A-pos.11	脚の近くに白抜け3カ所 three white spots near legs		
B-pos.17	尾の下に白点 white dot below tail		
B-pos.18	"5"に白点 white dot on "5"		
C-pos.1	背中に黒点 black dot on back		
C-pos.10	脚の下方に白点 white dot below leg		
D-pos.7	尾の上方に白点 white dot above tail		
D-pos.13	顎の下に白点4カ所 four white dots below chin		

N18 / N18A 1963年（昭和38）用 For 1963 (Showa 38)

発行日：N18=1962（昭和37）.12.15、N18A=1963（昭和38）.1.20
Date of Issue : N18=1962 (Showa 37).12.15, N18A=1963 (Showa 38).1.20

穿孔となった。
1) The perforation was changed to double comb with single hole for N18 and thereafter. Since the pane orientation on the plates of 20-subject pane was changed, it became possible to distinguish a stamp from 20-subject pane (short grain paper) from a stamp from souvenir sheet (long grain paper). The vertical dimension of the souvenir sheet was increased to 93.5 mm, and the perforation was changed to harrow type.

N18 のごみ人形のうさぎ・十二支土鈴（佐賀）
Nogomi Zodiac Bell Doll, Rabbit (Folk Toy of Saga)

N18A お年玉小型シート "New Year's Gift" Souvenir Sheet (lottery prize)

● 風景印 Picture Postmark

鹿島 Kashima

N18 ● 定常変種 Constant Flaws

A-pos.8 / A-pos.18 / B-pos.1 / B-pos.20

C-pos.1 / C-pos.19 / D-pos.1 / D-pos.2

	**	●	FDC
N18 5 yen 多色 multicolored	50	30	500
N18A 20 yen (5 yen × 4)	1,500	1,700	2,300

切手データ Basic Information

N18
発行枚数：1,500万枚　number issued : 15 million
製版・目打組合せ：タイプ2B-1　plate and perforation configuration : type 2B-1

N18A
発行枚数：1,804.7万枚　number issued : 18,047,000

位置 position	特徴 feature	**	●
A-pos.8	"5"の左方に白点 white dot to left of "5"	200	100
A-pos.18	"5"の左方に白点 white dot to left of "5"		
B-pos.1	胴体に紫点 purple dot on body		
B-pos.20	顔の左に白線 white line to left of face		
C-pos.1	胴体の左下に白点 white dot to lower left of body		
C-pos.19	"9"の上方に白点 white dot above "9"		
D-pos.1	笹に黒点 black dot on bamboo leaf		
D-pos.2	笹に黒点、笹の左に黒線 black dot on bamboo leaf, black line to left of bamboo leaf		

1) N18以降、目打穿孔はすべて連続目打の二連1型となった。また20枚シートの製版の向きが変更されたため、20枚シート（横紙）と小型シート（縦紙）が紙目によって識別できるようになった。小型シートの寸法は縦が拡大されて93.5㎜、目打は田型の全型目打により上段と下段の一括

N19 / N19A 1964年（昭和39）用 For 1964 (Showa 39)

発行日：N19=1963（昭和38）.12.16、N19A=1964（昭和39）.1.20
Date of Issue : N19=1963 (Showa 38).12.16, N19A=1964 (Showa 39).1.20

N19 挽物人形の辰（岩井）と木工品「龍神招福」（竜王）
Wood-turned Dragon (Iwai) and Wooden Toy "Ryujin-shofuku" (Ryuou)

N19A お年玉小型シート "New Year's Gift" Souvenir Sheet (lottery prize)

● 風景印 Picture Postmark

  岩井 Iwai   龍王 Ryuou

	**	●	FDC
N19 5 yen 多色 multicolored	40	30	500
N19A 20 yen (5 yen × 4)	800	900	1,500

切手データ Basic Information

N19
発行枚数：2,000万枚　number issued : 20 million
製版・目打組合せ：タイプ2B-1　plate and perforation configuration : type 2B-1

N19A
発行枚数：12,705,654枚　number issued : 12,705,654

●変則目打 Irregular Perforation

上耳紙 upper seivage

目打穴数 number of extension hole on upper selvage 0-1-0-1-0

目打穴数 number of extension hole on upper selvage 1-0-1-0-1

1）変則目打は、2版の一部シートに発見されている。上下の耳紙に目打穴が1個飛び出す二連1型で穿孔され、櫛歯端の縦目打上側の目打針が1本少なく植えられた。

1) An irregular perforation is known for some sheets plate 2. The perforation of this stamp was double comb type with one extension hole, and one top extension pin was omitted in the irregular perforation.

●スクリーン角度 Screen Angles

1版 Plate 1
緑56度
green 56 degrees

2版 plate 2
緑61度
green 61 degrees

N19 ●定常変種 Constant Flaws

| P1/P2-A-pos.15 | P1-A-pos.15 | P1/P2-B-pos.17 | P1-B-pos.8 |

P2-B-pos.3 | P1/P2-C-pos.1 | P1/P2-C-pos.16 | P1-C-pos.18

P2-C-pos.1 | P1/P2-D-pos.2 | P1-D-pos.4 | P2-D-pos.8

位置 position	特徴 feature	**	●
P1/P2-A-pos.15	尾の左に白点 white dot to left of tail	200	100
P1-A-pos.15	竜神の左に緑点 green dot to left of "Ryujin"		
P1/P2-B-pos.17	尾の左に太い赤線 thick red line to left of tail		
P1-B-pos.8	口の右に緑点 green dot to right of mouth		
P2-B-pos.3	緑の台に長い斜線 long diagonal green line on green board		
P1/P2-C-pos.1	竜神に赤点 red dot on "Ryujin"		
P1/P2-C-pos.16	右マージンに青点 blue dot in right margin		
P1-C-pos.18	"便"の右下方に緑線 green line to lower right of "便"		
P2-C-pos.1	竜神の右に緑点 green dot to right of "Ryujin"		
P1/P2-D-pos.2	竜神に濃い青点 dark blue dot on "Ryujin"		
P1-D-pos.4	右マージンに緑点 green dot in right margin		
P2-D-pos.8	竜神の右に緑点 green dot to right of "Ryujin"		

N20/N20A 1965年（昭和40）用　発行日：N20=1964（昭和39）.12.15、N20A=1965（昭和40）.1.20
For 1965 (Showa 40)　Date of Issue：N20=1964 (Showa 39).12.15, N20A=1965 (Showa 40).1.20

N20 わら細工の蛇（東京の郷土玩具）Straw Craft Snake (Folk Toy of Tokyo)

N20A お年玉小型シート "New Year's Gift" Souvenir Sheet (lottery prize)

切手データ Basic Information	
N20	
発行枚数：3,300万枚	number issued : 33 million
製版・目打組合せ：タイプ2B-1	plate and perforation configuration : type 2B-1
N20A	
発行枚数：15,325,973枚	number issued : 15,325,973

1）N20以降は、全色のスクリーン線数が250線、スクリーン角度が45度に統一された。

2）緑色の定常変種が確認できるものを1版、緑色の定常変種が確認できないものを2版に分類した。各刷色では、1版と2版に共通の定常変種は確認できていない。

1) The screen numbers and the screen angles of all colors were respectively unified to 250 and 45 degrees for N20 and thereafter.

2) The plate with constant flaws in the green plate is classified as plate 1, and that without constant flaws in the green plate is classified as plate 2. No common constant flaw of each color has been found.

●風景印 Picture Postmark

浅草 Asakusa

	**	●	FDC
N20 5 yen 多色 multicolored	40	30	350
N20A 20 yen (5 yen × 4)	450	500	1,200

年賀切手

N20 ●定常変種 Constant Flaws

P1-A-pos.19

P1-B-pos.15

P1-B-pos.17/18

P1-C-pos.16

P1-C-pos.19

P1-D-pos.12

P1-D-pos.20

P2-A-pos.2

P2-B-pos.8

P2-B-pos.10

P2-D-pos.16

P2-D-pos.18

位置 position	特徴 feature	**	●
P1-A-pos.19	"局"の上に緑点 green dot above "局"	200	100
P1-B-pos.15	"便"の下方に黒点 black dot below "便"		
P1-B-pos.17/18	マージンに赤線と赤点 rred line and red dot in margin between pos.17 and pos. 18		
P1-C-pos.16	左マージンに赤点 red dot in left margin		
P1-C-pos.19	"局"の右上に緑点 green dot to upper right of "局"		
P1-D-pos.12	第2コーナーに赤点 red dot in NE corner		
P1-D-pos.20	左マージンに黒点 black dot in left margin		
P2-A-pos.2	第1コーナーに赤点 red dot in NW corner		
P2-B-pos.8	上マージンに黒点 black dot in top margin		
P2-B-pos.10	"5"の中に赤点 red dot on "5"		
P2-D-pos.16	第3コーナーに赤点 red dot in SE corner		
P2-D-pos.18	"便"の左に黒点 black dot to left of "便"		

N21 / N21A 1966年（昭和41）用 For 1966 (Showa 41)

発行日：N21=1965（昭和40）.12.10、N21A=1966（昭和41）.1.20
Date of Issue：N21=1965 (Showa 40).12.10, N21A=1966 (Showa 41).1.20

N21 わら細工の忍び駒（岩手の郷土玩具）
Straw Craft Horse (Folk Toy of Iwate)

N21A お年玉小型シート
"New Year's Gift" Souvenir Sheet (lottery prize)

B-pos.18

C-pos.5

C-pos.12

D-pos.5

D-pos.16

D-pos.18

D-pos.6

● 風景印 Picture Postmark 　花巻 Hanamaki

	**	●	FDC
N21 5 yen 多色 multicolored	40	30	350
N21A 20 yen (5 yen × 4)	450	500	1,200

切手データ Basic Information

N21
発行枚数：3,500万枚　number issued : 35 million
製版・目打組合せ：タイプ2B-1　plate and perforation configuration : type 2B-1

N21A
発行枚数：1,688.9万枚　number issued : 16,889,000

N21 ●定常変種 Constant Flaws

A-pos.1

A-pos.9

B-pos.10

B-pos.12

位置 position	特徴 feature	**	●
A-pos.1	"便"の右下に赤点、尾に黒点 Red dot to lower right of "便, black dot on tail	200	100
A-pos.9	布に赤いシミ red stain on fabric		
B-pos.10	脚に大きな赤点 large red dot on leg		
B-pos.12	布に赤点 red dot on fabric		
B-pos.18	"6"に青い汚れ nick in "6"		
C-pos.5	脚に黒点 black dot on leg		
C-pos.12	"1"の左に濃い青いシミ dark blue stain to left of "1"		
D-pos.5	"5"の近くに白点3カ所 three white dots near "5"		
D-pos.6	脚に黒点と赤点 black dot and red dot on leg		
D-pos.16	"1"の左下方に白点 white dot to lower left of "1"		
D-pos.18	頭に赤点 red dot on head		

◆ 刷色とスクリーン角度 Printing Color and Screen Angle

番号 No.	刷色 printing color	スクリーン角度(°) screen angle (°)	
		20枚シート pane of 20	小型シート souvenir sheet
N8 / N8A	赤 red	46	45
N9 / N9A	赤 red	45	
N10 / N10A	赤／黒 red / black	45 / 2	
N11 / N11A	赤／緑 red / green	45 / 53	80 / 45
N12 / N12A	黒／青／赤／黄茶 black / blue / red / yellow-blown	63 / 88 / 62 / 45	68 / 1 / 62 / 45 72 / 1 / 57 / 45
N13 / N13A	黒／赤／緑／黄 black / red / green / yellow	45 / 72 / 55 / 86	45 / 68 / 55 / 86 45 / 73 / 62 / 90 45 / 75 / 55 / 86 58 / 68 / 55 / 86
N14 / N14A	赤／青／黄／黒 red / blue / yellow / black	60 / 56 / 84 / 45 65 / 56 / 84 / 45	57 / 53 / 89 / 45 75 / 56 / 89 / 45 75 / 61 / 89 / 45
N15 / N15A	黒／緑／赤／金 black / green / red / gold	57 / 62 / 45 / 79	53 / 68 / 45 / 2 57 / 68 / 45 / 2 58 / 62 / 45 / 89 58 / 73 / 45 / 90
N16 / N16A	金／赤／黄／緑 gold / red / yellow / green	87 / 60 / 68 / 45	1 / 45 / 78 / 45 1 / 55 / 78 / 45 5 / 45 / 75 / 45 5 / 60 / 65 / 45 5 / 60 / 75 / 45
N17 / N17A	黒／赤／黄／緑 black / red / yellow / green	54 / 90 / 71 / 45	57 / 3 / ? / 45 63 / 90 / 71 / 45
N18 / N18A	黒／橙／緑／紫 black / orange / green / purple	45 / 56 / 78 / 45	45 / 58 / 78 / 45 45 / 60 / 72 / 45 45 / 60 / 76 / 45 45 / 60 / 78 / 45
N19 / N19A	黒／赤／緑／金 black / red / green / gold	88 / 74 / 56 / 45 88 / 74 / 61 / 45	1 / 71 / 56 / 45 1 / 71 / 59 / 45 89 / 77 / 59 / 45 89 / 77 / 64 / 45

1) N20 以降は、全色 45 度に統一された。
1) From N20 and thereafter, the screen angles of all colors were unified to 45 degrees.

使用例評価 Valuations by Cover

※使用例評価は、適正（適応）使用を基準としている。
※ The value of covers are based on proper (adapted) usage.

● **N7 - N11 5 yen**
国内郵便 domestic mail
1枚貼私製はがき該当年の年賀状
new year's postcard for corresponding year bearing 5 yen stamp
　櫛型印（図入年賀印）comb type date stamp with new year's ·············· 30,000
　機械印（時刻表示"0-8"）machine date stamp with "0-8" ··········· 7,000
　1枚貼私製はがき postcard bearing 5 yen stamp ················· 800
　2枚貼書状 letter bearing two 5 yen stamps ··················· 1,000
　発行月内の非年賀使用 non-new year's mail within month of issue usage. ·· 10,000
　各種国内便 various domestic mail ··················· 1,000
外国郵便 international mail
　2枚貼外信印刷物 printed matter bearing two 5 yen stamps ····· 2,000
　3枚貼2倍重量外信印刷物 double weight printing matter bearing three 5 yen stamps ································ 3,000
　5枚貼特別地帯宛航空便書状 air mail letter to special zone bearing five 5 yen stamps ····························· 3,000

● **N12 - N21 5 yen**
国内郵便 domestic mail
1枚貼私製はがき該当年の年賀状
new year's postcard for corresponding year bearing 5 yen stamp
　櫛型印（C欄年賀印）comb type date stamp with new year's ··· 10,000
　機械印（年賀表示）machine date stamp with new year's ······ 3,000
　1枚貼私製はがき postcard bearing 5 yen stamp ················· 500
　2枚貼書状 letter bearing two 5 yen stamps ····················· 800
　発行月内の非年賀使用(N9-N11) non-new year's mail within month of issue usage. (N9-N11) ··················· 5,000
　各種国内便 various domestic mail ··················· 800
外国郵便 international mail
　3枚貼外信印刷物 printed matter bearing three 5 yen stamps ·· 2,000
　5枚貼特別地帯宛航空便書状 air mail letter to special zone bearing five 5 yen stamps ···································· 2,500
● **全貼小型シート Pasting of All Souvenir Sheet 20 yen**
　N7A ··· 35,000
　N8A ··· 25,000
　N9A-N10A ·· 15,000
　N11A ·· 10,000
　N12A ·· 6,000
　N13A-N18A ·· 3,000
　N19A ·· 2,000
　N20A-N21A ·· 1,000

東京中央　昭和 29.12.22
N10発行 3日目の葉書の一般郵便
Tokyo C.P.O. December.22.1954
Non New Year's Greeting postcard usage, postmarked on the third day of issue.

戦後に年賀状の特別取扱いが開始されたのは昭和24年年賀からで、1948年（昭和23年）12月15日から年賀特別取扱いが実施された。
年賀切手が前年12月に発行されるのは昭和28年12月25日発行の三春駒（N9）からで、翌年のN10〜N16が12月20日（N15のみ19日）、N17以降は15日の発行となっている。したがって、発行月内の非年賀使用（12月中に、年賀ではなく一般の郵便物に使用したもの）はFDCと郵趣家便を除けば、意外に少ない。

TACHIKAWA　1957.12.20　発行日に差出された在日米軍の軍事郵便局 APO323(立川) 宛葉書。在日米軍基地宛は外国郵便扱いであるが、料金は国内郵便料金と同じである。
TACHIKAWA December.20.1957　Postcard addressed to U.S. military post office in Japan, A.P.O. 323 in Tachikawa. Mail addressed to U.S. military bases in Japan is treated as international mail, but the rates are the same as for domestic mail.

年賀切手（消印別評価）

消印別評価 Valuations by Postmark

この項の評価は、切手の印面に消印が90％以上程度かかっている単片で、局名や年月日がわかるものを基準としている。局名の違いなどによる評価の違いは考慮していない。

The value in this section is based on postmarks of single stamps, for which the name of the post office and full date can be read as follows; Clear strikes of 90% image or more. Differences in value due to differences in post office names, etc. are not taken into account.

消印タイプ Postmark		和文印 Postmarks for Domestic Mail					欧文印 Postmarks for International Mail		
		局名左書 post office name written left to right							
		櫛型印 JCD			機械印 JMD		ローラー印 JRD	三日月型印 RSD	機械印 RMD	ローラー印 RRD
		時刻戦後型 JCD-TD	鉄道郵便 JCD-RW	年賀 JCD-NY	時刻戦後型 JMD-TD	年賀 JMD-NY				
カタログ番号 Catalogue No.	額面 Face Values	25.5.18	27.3.13	図3 fig.3					図1 fig.1	図2 fig.2
N7	5 yen	300	−	×	500	×	600	−	−	−
N8	5 yen	300	−	12,000	500	3,000	600	−	−	−
N9	5 yen	300	−	10,000	500	3,000	600	−	−	−
N10	5 yen	300	−	10,000	500	3,000	−	−	−	−
N11	5 yen	200	−	10,000	400	3,000	−	1,000	3,000	−
N12	5 yen	200	−	5,000	400	1,500	−	1,000	−	−
N13	5 yen	200	−	5,000	400	1,500	600	1,000	−	−
N14	5 yen	200	800	5,000	300	1,500	500	800	−	−
N15	5 yen	200	−	3,000	300	1,500	−	800	−	−
N16	5 yen	200	−	3,000	300	1,500	−	800	−	−
N17	5 yen	200	−	2,000	200	1,000	−	800	−	−
N18	5 yen	100	−	2,000	200	1,000	−	800	−	−
N19	5 yen	100	−	2,000	200	1,000	−	800	−	−
N20	5 yen	100	−	2,000	200	1,000	−	800	−	−
N21	5 yen	100	−	2,000	200	1,000	−	800	−	−

図1 fig.1 欧文機械印 RMD

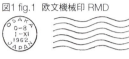

図2 fig.2 欧文ローラー印 RRD

図3 fig.3 櫛型年賀印 JCD-NY

 N8 N11

 N9 N12

 N10 N13-N21

凡例 Legend

Datestamp Type
<For Domestic Mail>
JCD: Comb Type Datestamps
JMD: Machine Datestamps
JRD: Roller Datestamps
<for International Mail>
RMD: Machine Datestamps
RSD: Swordguard Type Datestamps
RRD: Roller Datestamps

Postmark Type
JRD-1: post office name written horizontally left to right
JRD-2: post office name written vertically
NY: New Year postmarks
TD: postmarks with time display in 5 divisions
RW: railway post office postmarks

◆原画作者・原画構成者 Designer and Retoucher

番号 No.	原画作者 Designer	原図修正者 Retoucher
N7	吉田豊 Yoshida Yutaka	綱島廉 Tsunajima Ren
N8	加曾利鼎造 Kasori Teizo	前野京平 Maeno Kyohei
N9	木村勝 Kimura Masaru	島田武夫 Shimada Takeo
N10	渡辺三郎 Watanabe Saburo	山野内孝夫 Yamanouchi Takao
N11	久野実 Hisano Minoru	綱島廉 Tsunajima Ren
N12	長谷部日出男 Hasebe Hideo	綱島廉 Tsunajima Ren
N13	木村勝 Kimura Masaru	丹後昭 Tango Akira
N14	木村勝 Kimura Masaru	前野京平 Maeno Kyohei
N15	木村勝 Kimura Masaru	東角井良臣・山田輝郎 Higashitsunoi Yoshiomi, Yamada Teruo
N16	木村勝 Kimura Masaru	丹後昭 Tango Akira
N17	江守若菜 Emori Wakana	前野京平 Maeno Kyohei
N18	江守若菜 Emori Wakana	前野京平 Maeno Kyohei
N19	木村勝 Kimura Masaru	不明 unknown
N20	日置勝駿 Hioki Masatoshi	不明 unknown
N21	江守若菜 Emori Wakana	不明 unknown

Postal Stationery ステーショナリー

記念・特殊はがき Commemorative and Special Postal Cards

記念・特殊はがきとは料額印面付きの官製はがきで、料額印面のない記念絵はがき等は含まない。1959年（昭和34）からは、「敬老の日」や「成人の日」などの特殊はがきが1975年（昭和50）まで毎年発行された。

Commemorative & special postal cards are official postal cards with an imprinted stamp, and do not include commemorative and special picture postcards without the imprinted stamp. From 1959 (Showa 34), special postal cards such as for "Respect for the Aged Day" and "Coming of Age Day" were issued every year until 1975 (Showa 50).

●共通データ Common Information
版式：CC7、CC9〜CC22＝平版、CC8＝グラビア版、サイズ：90×140mm、用紙：CC7＝灰白紙、CC8〜CC22＝白色洋紙
Printing：CC7, CC9-CC22=Offset Printing, CC8=Photogravure Printing, Size：90×140mm,
Paper: CC7=Greyish Paper, CC8-CC22=White Foreign Paper

CC7 平和条約発効・憲法5周年 5th Anniversary of Peace Treaty and Constitution
発行日：1952（昭和27）.5.3
Date of Issue：1952 (Showa 27).5.3
CC7 アヤメ、シャクヤク、バラ
Iris, Chinese Peony and Rose
CC7 5 yen にぶ赤紫 reddish purple (500,000) ** ● FDC
………………………… 800　600　1,000

CC8 印刷文化典 Printing Culture Exhibition
発行日：1952（昭和27）.9.15
Date of Issue：1952 (Showa 27).9.15
CC8 日本最初の印刷機
Japan's First Printing Press
CC8 5 yen 緑 green (500,000)
………………………… 1,400　600　1,000

CC9 母の日・こどもの日 Mother's Day and Children's Day
発行日：1958（昭和33）.5.1
Date of Issue：1958 (Showa 33).5.1
CC9 こいのぼり Carp Streamer
CC9 5 yen こい赤・暗い青緑 red and dark blue-green (2,000,000)
………………………… 600　400　1,000

CC10 敬老の日 Respect for the Aged Day
発行日：1958（昭和33）.9.13
Date of Issue：1958 (Showa 33).9.13

CC10 老人と子供 Senior Citizen and Child
CC10 5 yen こい赤茶紫 deep mauve (5,000,000) ……… 500　300　600

CC11 成人の日 Coming of Age Day
発行日：1959（昭和34）.1.12
Date of Issue：1959 (Showa 34).1.12
CC11 若人の象徴 Allegory of Youth
CC11 5 yen 暗い青緑 dark blue-green ** ● FDC
(5,000,000) ……… 500　300　600

CC12 国土緑化運動 National Land Afforestation Campaign
発行日：1959（昭和34）.4.3
Date of Issue：1959 (Showa 34).4.3
CC12 樹木 Tree
CC12 5 yen 緑・黒 green and black (5,000,000) ……… 500　300　600

CC13 母の日・こどもの日 Mother's Day and Children's Day
発行日：1959（昭和34）.5.1
Date of Issue：1959 (Showa 34).5.1
CC13 鳥の親子 Bird Family
CC13 5 yen 赤・暗緑 red and dark green (10,000,000) ……… 500　300　600

CC14 敬老の日 Respect for the Aged Day
発行日：1959（昭和34）.9.12
Date of Issue：1959 (Showa 34).9.12
CC14 おもと Rohdea Japonica
CC14 5 yen こい赤紫・緑 red-purple and green (5,000,000) ……… 500　300　600

ステーショナリー（記念・特殊はがき／年賀はがき）

CC15 成人の日 Coming of Age Day

発行日：1960（昭和35）.1.11
Date of Issue：1960 (Showa 35).1.11

CC15 寒梅に雪 Winter Plum and Snow

	**	●	FDC
CC15 5 yen にぶ赤・黄緑・うす青 red, green and blue (5,000,000)	500	300	600

CC16 特許制度75年 75th Anniversary of Patent System

発行日：1960（昭和35）.4.18
Date of Issue：1960 (Showa 35).4.18

CC16 発明の象徴 Allegory of Invention

CC16 5 yen 灰味青・茶黒 grayish blue and black (5,000,000) …… 500 300 600

CC17 母の日・こどもの日 Mother's Day and Children's Day

発行日：1960（昭和35）.4.25
Date of Issue：1960 (Showa 35).4.25

CC17 遊ぶ母子 Mother and Child at Play

CC17 5 yen 赤紫・緑 reddish purple and green (5,000,000) …… 400 250 600

CC18 敬老の日 Respect for the Aged Day

発行日：1960（昭和35）.9.1
Date of Issue：1960 (Showa 35).9.1

CC18 かぐや姫 Princess Kaguya

CC18 5 yen こい緑・黄味だいだい dark green and orange (5,000,000) …… 400 250 600

CC19 成人の日 Coming of Age Day

発行日：1961（昭和36）.1.10
Date of Issue：1961 (Showa 36).1.10

CC19 青年男女 Young Man and Woman

	**	●	FDC
CC19 5 yen うす赤・暗い茶 pale red and dark brown (5,000,000)	400	250	600

CC20 母の日・こどもの日 Mother's Day and Children's Day

発行日：1961（昭和36）.5.1
Date of Issue：1961 (Showa 36).5.1

CC20 カーネーション Carnation

CC20 5 yen 赤・にぶ青緑・黒 red, blue-green and black (5,000,000) …… 400 250 600

CC21 敬老の日 Respect for the Aged Day

発行日：1961（昭和36）.9.1
Date of Issue：1961 (Showa 36).9.1

CC21 末広 Japanese Fan

CC21 5 yen 灰味茶・青味緑 greyish brown and blue-green (5,000,000) ・400 250 450

CC22 成人の日 Coming of Age Day

発行日：1962（昭和37）.1.10
Date of Issue：1962 (Showa 37).1.10

CC22 若駒 Young Horses

CC22 5 yen 茶・黄味緑 brown and yellow-green (5,000,000) ……… 400 250 450

年賀はがき New Year's Greeting Postal Cards

「お年玉つき年賀はがき」としてくじ付きの官製年賀はがきが、1949年（昭和24）に法令に基づいて初めて発行された。1952年（昭和27）用は寄附金付きの1種を発行、1953年（昭和28）用からは寄附金なしと寄附金付きの2種が発行された。1962年（昭和37）用からは、官製年賀はがきによる年賀状には引受消印を省略することになり、料額印面の下部に消印図案が印刷された。

"New Year's Greeting Postal Cards" with a lottery number were first issued pursuant to a new law in 1949 (Showa 24). In 1952 (Showa 27), one type of postal card with donation (semi-postal) was issued. From 1953 (Showa 28), two types of postal cards were issued: one with donation and one without donation. From 1962 (Showa 37), postmarks were omitted for New Year's Greeting Postal Cards, as the postmark was printed as part of the design.

● 共通データ（特記外）Common Information
版式：NC4〜NC12、NC17〜NC30＝平版、NC13〜NC16＝凸版、サイズ：90×140mm、用紙：淡色洋紙
Printing: NC4-NC12, NC17-NC30=Offset Printing, NC13-NC16=Relief Printing, Size: 90×140mm,
Paper: Tinted Foreign Paper

NC4 1952年（昭和27）用 For 1952 (Showa 27)

発行日：1951（昭和26）.11.15
Date of Issue：1951 (Showa 26).11.15

NC4 やっこだこ Kite Flying

	*	●
NC4 2 yen + 1 yen 黄味赤 yellow-red (350,000,000)	1,500	200
a. 銘版なし without imprint	1,600	300

ステーショナリー（年賀はがき）

NC4 銘版あり
(印刷庁製造)
with imprint
組番号（001〜170）
group No.001-170

NC4a 銘版なし
without imprint
組番号（171〜350）
group No.171-350

1) NC4からNC20までの使用済評価は年賀機械印が標準で、年賀手押し印の場合は評価が2倍となる。
1) The prices for used copies from NC4 to NC20 are for postal cards cancelled with machine New Year postmark. The prices of the postal cards cancelled with hand-struck New Year postmark are double of those prices.

NC5-6　1953年（昭和28）用 For 1953 (Showa 28)

発行日：1952（昭和27）.11.15
Date of Issue：1952 (Showa 27).11.15

（左）NC5 鶴と日の出
(left) Crane and Sunrise
（右）NC6 梅花模様
(right) Plum Pattern

NC5 4 yen こい茶 dark brown (100,000,000) くじ番号なし
....5,500　300
NC6 4 yen+1 yen 紅 deep red (350,000,000)......1,000　150

NC7-8　1954年（昭和29）用 For 1954 (Showa 29)

発行日：1953（昭和28）.11.15
Date of Issue：1953 (Showa 28).11.15

（左）NC7 鶴と日の出
(left) Crane and Sunrise
（右）NC8 末広
(right) Japanese Fan

NC7 4 yen 赤味茶 red-brown (100,000,000) くじ番号なし
....5,500　200
NC8 4 yen + 1 yen 紅赤 red (400,000,000)......1,000　100

1) NC7の印面はNC5と同様で、宛名面下部に「差出上の注意」が印刷されている。またNC7は、1955年（昭和30）用、1956年（昭和31）用として各1億枚追加発行された。
1) The imprinted stamp of NC7 is same as that of NC5, but "mailing instructions" are printed at the bottom of the address side. One hundred million additional copies of NC7 were printed for both 1955 (Showa 30) and 1956 (Showa 31) respectively.

NC9　1955年（昭和30）用 For 1955 (Showa 30) Coming of Age Day

発行日：1954（昭和29）.11.15
Date of Issue：1954 (Showa 29).11.15

NC9 つづみ Tsuzumi (hand drum)

NC9 4 yen + 1 yen 紅 deep red (480,000,000)
....800　50

NC10　1956年（昭和31）用 For 1956 (Showa 31)

発行日：1955（昭和30）.11.15
Date of Issue：1955 (Showa 30).11.15

NC10 福寿草 Amur Adonis

NC10 4 yen + 1 yen こい紫味紅 purplish red (520,000,000)...................................800　50

NC11-12　1957年（昭和32）用 For 1957 (Showa 32)

発行日：1956（昭和31）.11.15
Date of Issue：1956 (Showa 31).11.15

（左）NC11 折り鶴
(left) Origami Crane
（右）NC12 松竹梅
(right) Pine, Bamboo and Plum

NC11 4 yen 赤味だいだい red-orange (300,000,000)
....400　50
NC12 4 yen + 1 yen 赤 red (450,000,000)............600　50

1) この年から寄附金なしにもお年玉くじが付いた。
1) Starting this year, a lottery number was included also on the New Year's Greeting Postal Cards without donation.

NC13-14　1958年（昭和33）用 For 1958 (Showa 33)

発行日：1957（昭和32）.11.15 Date of Issue：1957 (Showa 32).11.15

（左）NC13 富士山
(left) Mt.Fuji
（右）NC14 梅にウグイス
(right) Plum and Warbler

NC13 4 yen 赤 red (250,000,000) ....500　50
NC14 4 yen + 1 yen こい紫味赤 purplish red (450,000,000)
....600　50

NC15-16　1959年（昭和34）用 For 1959 (Showa 34)

発行日：1958（昭和33）.11.15 Date of Issue：1958 (Showa 33).11.15

（左）NC15 ナンテン
(left) Nandina Domestica
（右）NC16 伊勢海老
(right) Japanese Lobster

NC15 4 yen こい紫味赤 purplish red (170,000,000)
....500　50
NC16 4 yen + 1 yen 赤味だいだい red-orange (600,000,000)
....600　50

ステーショナリー（年賀はがき）

NC17-18　1960年（昭和35）用 For 1960 (Showa 35)

発行日：1959（昭和34）.11.15 Date of Issue：1959 (Showa 34).11.15

（左）NC17 梅花模様
(left) Plum Pattern

（右）NC18 初日の出
(right) First Sunrise of Year

NC17　4 yen 紫 purple (170,000,000) ………… 400　50
NC18　4 yen + 1 yen 赤味だいだい red-orange (620,000,000)
　　　　　　　　　　　　　　　　………… 600　50

NC19-20　1961年（昭和36）用 For 1961 (Showa 36)

発行日：1960（昭和35）.11.15 Date of Issue：1960 (Showa 35).11.15

（左）NC19 万歳　(left) Banzai (Japanese Traditional Entertainment)

（右）NC20 振りづち (right) Mallet (Symbol of Good Fortune)

NC19　4 yen 暗い赤 dark red (220,000,000) ……… 400　50
NC20　4 yen + 1 yen 赤 red (630,000,000) ………… 450　50

NC21-22　1962年（昭和37）用 For 1962 (Showa 37)

発行日：1961（昭和36）.11.15 Date of Issue：1961 (Showa 36).11.15

（左）NC21 しめ飾り
(left) Japanese New Year's Wreath

（右）NC22 鶴と松
(right) Crane and Pine

NC21　4 yen 赤紫 reddish purple (230,000,000)
　　　　　　　　　　　　　………… 350　30
NC22　4 yen + 1 yen 赤 red (640,000,000)… 400　30

1）この年から料額印面の下部に消印図案が付いた。
1) From this year, the postmark was printed below the imprinted stamp as part of the design.

NC23-24　1963年（昭和38）用 For 1963 (Showa 38)

発行日：1962（昭和37）.11.15 Date of Issue：1962 (Showa 37).11.15

（左）NC23 鯛
(left) Sea Bream

（右）NC24 日の出と船
(right) Sunrise and Ship

NC23　4 yen 赤味だいだい red-orange (260,000,000)
　　　　　　　　　　　　　………… 350　30
NC24　4 yen + 1 yen 暗い赤紫 dark reddish purple (610,000,000)
　　　　　　　　　　　　　………… 400　30

NC25-26　1964年（昭和39）用 For 1964 (Showa 39)

発行日：1963（昭和38）.11.15 Date of Issue：1963 (Showa 38).11.15

（左）NC25 寒牡丹
(left) Winter Peony

（右）NC26 凧
(right) Kite

　　　　　　　　　　　　　　　　　　　　* ●
NC25　4 yen 黄味赤 yellowish red (330,000,000) ··350　30
NC26　4 yen + 1 yen こい赤紫 deep reddish purple
　　　　(610,000,000) ………………………… 400　30

NC27-28　1965年（昭和40）用 For 1965 (Showa 40)

発行日：1964（昭和39）.11.12 Date of Issue：1964 (Showa 39).11.12

（左）NC27 三宝みかん
(left) Sanbokan (Sour Orange)

（右）NC28 とそのちょうし
(right) Decanter of New Year's Spiced Sake

NC27　4 yen 黄味赤 yellow-red (475,000,000) ……… 350　30
NC28　4 yen + 1 yen こい赤紫 deep reddish purple
　　　　(610,000,000) ………………………… 400　30

NC29-30　1966年（昭和41）用 For 1966 (Showa 41)

発行日：1965（昭和40）.11.10 Date of Issue：1965 (Showa 40).11.10

（左）NC29 高砂
(left) Elder Couple (Legendary Sweethearts)

（右）NC30 鏡餅
(right) Japanese New Year Rice Cake

NC29　4 yen 黄味赤 yellow-red (560,000,000) ……… 350　30
NC30　4 yen + 1 yen こい赤紫 deep red-purple (650,000,000)
　　　　　　　　　　　　　………………………… 400　30

季節見舞はがき Seasonal Greeting Postal Cards

暑中見舞はがき
Summer Greeting Postal Cards

年賀はがきに続く郵政事業の増収策の一環として、1950年（昭和30）から官製による絵入りの暑中見舞はがきが発行され、1952年（昭和32）からは季節見舞はがき独自の印面図案となった。

As part of a strategy to increase revenue for the postal service following the New Year's Greeting Postal Cards, illustrated summer greeting cards were issued by the government from 1950 (Showa 30). Seasonal Greeting Postal Cards with imprinted stamps unique to the series were issued beginning in 1952 (Showa 32).

裏面図案（例）
Back Design (example)

● 共通データ（特記外）Common Information
版式：平版、サイズ：90×140mm、用紙：白色洋紙 Printing: Offset Printing, Size: 90×140mm, Paper: White Foreign Paper
1) 色調の後の（ ）内の数字は発行枚数、裏面図案は2種で発行枚数は2種とも同じ。1) The number in parentheses ( ) after the color shade is the quantity issued. There are two types of designs on the back, and the quantity issued is the same for both types.

SG3 波模様
Wave Pattern

SG4 スイレン
Water Lily

SG5 うちわ
Japanese Fan

SG6 ホタル
Firefly

SG7 貝
Shellfish

SG8 魚
Fish

SG9 トンボ
Dragonfly

SG10 熱帯魚
Tropical Fish

SG11 ホテイアオイ
Water Hyacinth

SG12 蝉
Cicada

SG13 夜顔
Moonflower

SG14 ナナホシテントウ
Seven-spot Ladybird

SG15 花火
Fireworks

SG16 すだれとうちわ
Blind and Fan

1952（昭和27）.7.1　　　　　　　　　　　　　　　＊　　●
SG3　5 yen 暗い緑 dark green (19,368,400)………5,000　800

1953（昭和28）.7.10
SG4　5 yen 暗い青 dark blue (20,000,000)………4,000　400

1954（昭和29）.7.1
SG5　5 yen 青緑 blue-green (20,000,000)…………3,000　600

1955（昭和30）.7.1
SG6　5 yen にぶ緑 green (30,000,000)……………2,500　300

1956（昭和31）.7.1
SG7　5 yen 暗い青 dark blue (30,000,000)………2,300　300

1957（昭和32）.7.1
SG8　5 yen 暗い青 dark blue (30,000,000)………1,500　300

1958（昭和33）.7.1
SG9　5 yen 暗い青 dark blue (30,000,000)………1,300　200

1959（昭和34）.7.1　　　　　　　　　　　　　　　＊　　●
SG10　5 yen 暗い青・青 dark blue (30,000,000)‥1,200　200

1960（昭和35）.7.1
SG11　5 yen 暗い青緑、うす青 blue-green (30,000,000)‥800　200

1961（昭和36）.7.1
SG12　5 yen 緑・黒 green and black (30,000,000)…700　150

1962（昭和37）.7.2
SG13　5 yenにぶ緑・暗い青 green and blue (30,000,000)・700　150

1963（昭和38）.7.1
SG14　5 yen 赤・濃い青 red and blue (30,000,000)…600　150

1964（昭和39）.7.1
SG15　5 yenうす赤・赤味青 red and blue (40,000,000)…500　150

1965（昭和40）.7.1
SG16　5 yen 暗い青緑・黄味赤 blue-green and red (45,000,000)
………………………………………………………350　150

製版・目打組合せ型式 Types of Plate and Perforation Configuration

型式 types		実用版構成 printing plate configuration	目打穿孔 pane configuration for perforations	目打の抜け方 position of perforations through selvages
			グループA：普通櫛型目打 Group A：Normal comb perforation	
1A-1		2面頭合せ 2-pane head-to-head configuration	2面縦並び 2-pane vertical configuration	右抜け・左抜け right or left selvage
1A-2		2×2面頭合せ 2×2-pane head-to-head configuration		
1A-3			2面横並び 2-pane horizontal configuration	下抜け bottom selvage
1A-4				上抜け top selvage
1A-5		3×2面頭合せ 3×2-pane head-to-head configuration	2面縦並び 2-pane vertical configuration	右抜け・左抜け right or left selvage
1B-1		2×2面順並び 2×2-pane normal configuration	2面縦並び 2-pane vertical configuration	右抜け right selvage
1B-2				左抜け left selvage
1B-3			2面横並び 2-pane horizontal configuration	下抜け bottom selvage
1B-4				上抜け top selvage
1C-1		2×2面順並び 2×2-pane normal configuration	4面 2×2-pane configuration	右抜け・左右抜け right selvage or both side selvages
1C-2				左抜け・左右抜け left selvage or both side selvages
1C-3				下抜け・上下抜け bottom selvage or top-and-bottom selvages
1C-4				上抜け・上下抜け top selvage or top-and-bottom selvages
1C-5				上下抜け top and bottom selvages
1D-1		2×2面順並び 2×2-pane normal configuration	2面縦並び 2-pane vertical configuration	下抜け・上下抜け bottom selvage or top-and-bottom selvages
1D-2				上抜け・上下抜け top selvage or top-and-bottom selvages
1D-3				上下抜け top and bottom selvages
1D-4			2面横並び 2-pane horizontal configuration	左抜け・左右抜け left selvage or both side selvages
1D-5				右抜け・左右抜け right selvage or both side selvages
1E-1		2×2面頭合せ 2×2-panes head-to-head configuration	1面 single pane	下抜け bottom selvage
1E-2				左抜け left selvage
1E-3		2×2面順並び 2×2-pane normal configuration		下抜け bottom selvage
1F		6面縦並び6-pane vertical configuration	3面縦続き 3-pane vertical configuration	上下抜け top and bottom selvages
1G		3面縦並び 3-pane vertical configuration	1面 single pane	下抜け bottom selvage
1H		2面横合せ 2-pane horizontal configuration	2面横続き 2-pane horizontal configuration	左抜け・左右抜け left selvage or both side selvages
1J-1		2面頭合せ 2-pane head-to-head configuration	1面 single pane	右抜け right selvage
1J-2				左抜け left selvage
1J-3				下抜け bottom selvage
1J-4				上抜け top selvage
1K-1		2×2面頭合せ 2×2-pane head-to-head configuration	2面横続き 2-pane horizontal configuration	左抜け・左右抜け left selvage or both side selvages
1K-2				右抜け・左右抜け right selvage or both side selvages

凡例 Note
- 縦型切手 vertical stamps
- 横型切手 horizontal stamps

※型式の**太字**は本カタログ掲載 The types in bold are listed in this catalogue.

●type1A

1A-2 横型切手・2×2面頭合せ、2面縦並び穿孔、右抜け・左抜け
horizontal stamps, 2×2-pane head-to-head configuration, 2-pane vertical configuration, right or left selvage

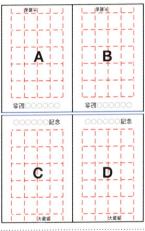

1A-3 縦型切手・2×2面頭合せ、2面横並び穿孔、下抜け
vertical stamps, 2×2-pane head-to-head configuration, 2-pane horizontal configuration, bottom selvage

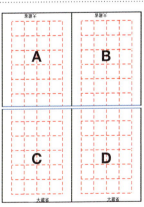

1A-4 縦型切手・2×2面頭合せ、2面横並び穿孔、上抜け
vertical stamps, 2×2-pane head-to-head configuration, 2-pane horizontal configuration, top selvage

●type1B

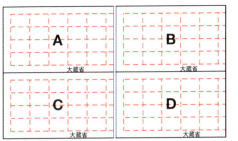

1B-1 横型切手・2×2面順並び、2面縦並び穿孔、右抜け
horizontal stamps, 2×2-pane normal configuration, 2-pane vertical configuration, right selvage

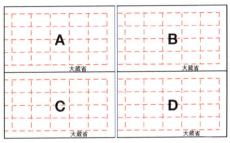

1B-2 横型／縦型切手・2×2面順並び、2面縦並び穿孔、左抜け
horizontal／vertical stamps, 2×2-pane normal configuration, 2-pane vertical configuration, left selvage

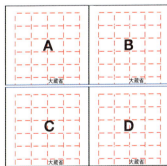

1B-3 横型／縦型切手・2×2面順並び、2面横並び穿孔、下抜け
horizontal／vertical stamps, 2×2-pane normal configuration, 2-pane vertical configuration, right selvage

1B-4 縦型切手・2×2面順並び、2面横並び穿孔、上抜け
horizontal stamps, 2×2-pane normal configuration, 2-pane vertical configuration, right selvage

実用版構成の2×2面が「頭合せ」から「順並び」に切り替わったのは、1958年8月発行のP200-201（佐渡国定公園）である。1958年3月～6月発行のC272（関門トンネル）とC274～C279（日本開港～ブラジル移住）は「頭合せ」としているが確証なく、「順並び」の可能性もある。
The 2×2-pane plate configuration changed from "head-to-head" to "normal" with P200 – P201, issued in August 1958. The plate configurations of C272 and C274 – C279, issued March – June 1958, are described as "head-to-head", but there is no definite proof. It is possible that they are "normal".

●type1C

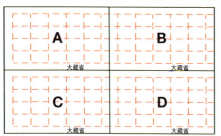

1C-1
横型／縦型切手・2×2面順並び、4面穿孔、右抜け・左右抜け
horizontal／vertical stamps, 2×2-pane normal configuration, 2×2-pane configuration, right selvage or both side selvages

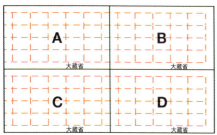

1C-2
横型／縦型切手・2×2面順並び、4面穿孔、左抜け・左右抜け
horizontal／vertical stamps, 2×2-pane normal configuration, 2×2-pane configuration, left selvage or both side selvages

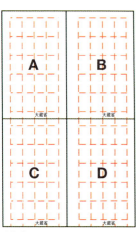

1C-3
縦型／横型切手・2×2面順並び、4面穿孔、下抜け・上下抜け
vertical／horizontal stamps, 2×2-pane normal configuration, 2×2-pane configuration, bottom selvage or top-and-bottom selvages

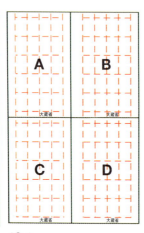

1C-4
縦型／横型切手・2×2面順並び、4面穿孔、上抜け・上下抜け
vertical／horizontal stamps, 2×2-pane normal configuration, 2×2-pane configuration, top selvage or top-and-bottom selvages

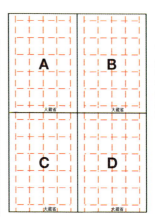

1C-5
縦型切手・2×2面順並び、4面穿孔、上下抜け
vertical stamps, 2×2-pane normal configuration, 2×2-pane configuration, top-and-bottom selvages

●type1D

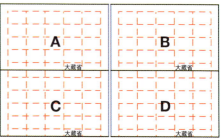

1D-1
横型切手・2×2面順並び、2面縦並び穿孔、下抜け・上下抜け
horizontal stamps, 2×2-pane normal configuration, 2-pane vertical configuration, bottom selvage or top-and-bottom selvages

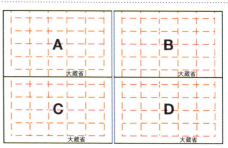

1D-2
横型切手・2×2面順並び、2面縦並び穿孔、上抜け・上下抜け
horizontal stamps, 2×2-pane normal configuration, 2-pane vertical configuration, top selvage or top-and-bottom selvages

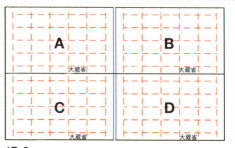

1D-3
横型切手・2×2面順並び、2面縦並び穿孔、上下抜け
horizontal stamps, 2×2-pane normal configuration, 2-pane vertical configuration, top-and-bottom selvages

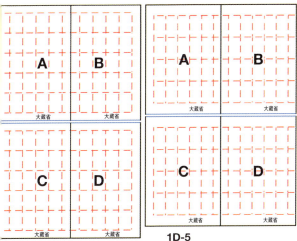

1D-4
縦型切手・2×2面順並び、2面横並び穿孔、左抜け・左右抜け
vertical stamps, 2×2-pane normal configuration, 2-pane horizontal configuration, left selvage or both side selvage

1D-5
縦型切手・2×2面順並び、2面横並び穿孔、右抜け・左右抜け
vertical stamps, 2×2-pane normal configuration, 2-pane horizontal configuration, right selvage or both side selvages

●type1E

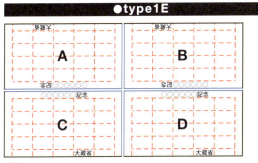

1E-1
横型切手・2×2面頭合せ、1面穿孔、下抜け
horizontal stamps, 2×2-pane head-to-head configuration, single pane, bottom selvage

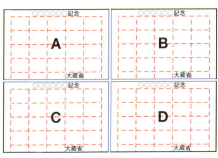

1E-3
横型切手・2×2面順並び、1面穿孔、下抜け
horizontal stamps, 2×2-pane normal configuration, single pane, bottom selvage

●type1F～1K

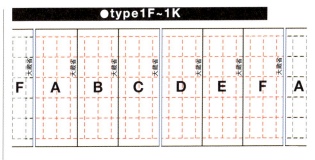

1F
横型切手・6面縦並び、3面縦続き穿孔、上下抜け
horizontal stamps, 6-pane vertical configuration, 3-pane vertical configuration, top and bottom selvages

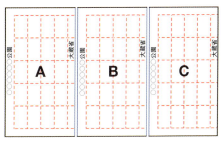

1G
横型切手・3面縦並び、1面穿孔、下抜け
horizontal stamps, 3-pane vertical configuration, single pane, bottom selvage

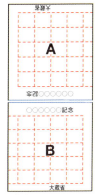

1J-2
横型切手・2面頭合せ、1面穿孔、左抜け
horizontal stamps, 2-pane head-to-head configuration, single pane, left selvage

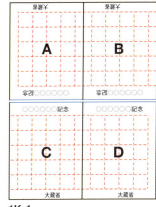

1K-1
縦型切手・2×2面頭合せ、2面横続き穿孔、左抜け・左右抜け
vertical stamps, 2×2-pane head-to-head configuration, 2-pane horizontal configuration, left selvage or both side selvages

巻末資料

型式 types	実用版構成 printing plate configuration	目打穿孔 pane configuration for perforations	目打の抜け方 position of perforations through selvages
	グループB：連続櫛型目打　Group B：Continuous comb perforation		
櫛型目打　Comb perforation			
2A	2面縦並び 2-pane vertical configuration	縦連続 horizontal configuration web paper	上下抜け top and bottom selvages
2B-1	4面横並び 4-pane horizontal configuration	横連続 horizontal configuration web paper	左右抜け left and right selvages
2B-2	4面縦並び 4-pane vertical configuration	縦連続 vertical configuration web paper	上下抜け top and bottom selvages
2C-1	3面横並び 3-pane vertical configuration	横連続 horizontal configuration web paper	左右抜け left and right selvages
2C-2	3面縦並び 3-pane vertical configuration	縦連続 vertical configuration web paper	上下抜け top and bottom selvages
全型目打　Harrow perforation			
2D	6面縦並び 6-pane vertical configuration	縦連続 vertical configuration web paper	1シート毎 one per sheet

凡例 Note

 縦型切手 vertical stamps

 横型切手 horizontal stamps

※型式の**太字**は本カタログ掲載 The types in bold are listed in this catalogue.

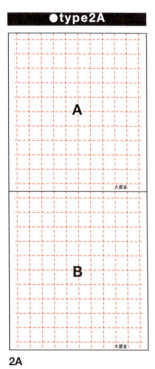

2A
縦型切手・2面縦並び、縦連続穿孔、上下抜け
vertical stamps, 2-pane vertical configuration, vertical configuration web paper, top and bottom selvages

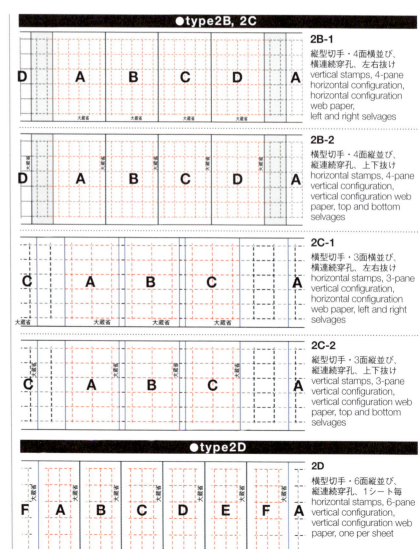

2B-1
縦型切手・4面横並び、横連続穿孔、左右抜け
vertical stamps, 4-pane horizontal configuration, horizontal configuration web paper, left and right selvages

2B-2
横型切手・4面横並び、縦連続穿孔、上下抜け
horizontal stamps, 4-pane vertical configuration, vertical configuration web paper, top and bottom selvages

2C-1
横型切手・3面横並び、横連続穿孔、左右抜け
horizontal stamps, 3-pane vertical configuration, horizontal configuration web paper, left and right selvages

2C-2
縦型切手・3面縦並び、縦連続穿孔、上下抜け
vertical stamps, 3-pane vertical configuration, vertical configuration web paper, top and bottom selvages

2D
横型切手・6面縦並び、縦連続穿孔、1シート毎
horizontal stamps, 6-pane vertical configuration, vertical configuration web paper, one per sheet

ご当地局の定義と類型 Definition and Types of Related Area Post Office

■ご当地局の基本定義

ご当地局(使用)とは、発行意図や図柄に地域性がある切手において、当該地域郵便局として郵政的根拠を有する郵便局及びその局による使用例である。以前から一部の収集家に好まれてきたものだが、これまでに明確に定義されたことはなく体系的な切手展作品も発表されてはいない。

本カタログでは初めて定義し、ご当地局を8つに類型化し、それぞれの切手についてご当地局を一覧で示した(p157-160)。FDC初日印局以外の特印使用局、関連小型印使用局、切手発行に合わせて風景印の図柄を刷新した局、イベント会場内臨時局・出張所などに注目することで、多様なご当地収集の世界を楽しむことができる。なお地域性がない(ご当地局がない)切手も存在することや、タイプ⑦についてはすべてをリスト化することは困難であるため、代表的な局を例示した。

リストには「ご当地局」×「消印種」において、確認数が少なく希少性があると思われるものに「★」を付けた(タイプ⑥⑦⑧は櫛型印を念頭に印を付けた)。FDCについても本リストをもとにチェックしてみると、新たな価値を見出すことができるだろう。

ご当地使用という発想を取り入れることで、使用例収集の幅が広がる。また、ご当地局の消印だけでなく、差出人や受取人、使用した封筒やはがきなどの地域性にも目を向けるとストーリー性が生まれて、さらに収集が豊かになる。記念押印から実逓へ郵趣家便から非郵趣家便へとこだわれば、難易度は一気に高くなるものの奥の深い収集になる。

■Definition of Related Area Post Office

A Related Area Post Office (gotōchi yūbinkyoku) is a post office recognized by the postal administration as having a special local relationship to a stamp with a regional design or regional purpose of issue. Usages from such post offices are characterized as "Related Area Post Office usage". Although examples of this kind of usage have been popular among some philatelists, it has never been clearly defined, nor has there been any philatelic event in which Related Area Post Office usage material has been systematically exhibited.

The catalogue is the first to provide a definition for Related Area Post Office usage, classifying it into eight types (pp.157-160). The Related Area Post Office for each stamp is shown in the following table. Changing the focus of attention from first-day datestamps of designated FDC post offices (type ②), produced by a private FDC provider, to the Related Area Post Offices that prepared their own commemorative postmarks or created new pictorial postmarks, and temporary post offices set up for events connected with the stamps, offers collectors the opportunity to enjoy a broad and diverse field of specialization. There are of course stamps without a related region, and such stamps do not have a Related Area Post Office. As it is difficult to list all stamps classified in type ⑦, selected examples from representative post offices are listed for type ⑦. Combinations of the Related Area Post Offices and types of postmarks are listed in the table, and scarce combinations are marked with a "★". (For types ⑥, ⑦, and ⑧, comb type datestamps are thought of as "★".) FDC's can take on additional value when examined against this list. Incorporating the concept of Related Area Post Office usage can expand the scope of collecting by usage type. Considering the regionality of senders, recipients, envelopes, and post cards can enrich a collection by adding a narrative aspect. Changing the focus of collecting from commemoratively made usages to actual usages, and from philatelic covers to the non-philatelic, will greatly increase the challenges but also add significant depth to a collection.

■ご当地局の類型 Types of Related Area Post Office

Type❶ 初日指定局 First Day Postmark designated Post Offices

1965年9月21日より初日通信日付印が開始になり、初日指定局が発表されるようになった。地域性を持つ切手の初日指定局はご当地局。

C436-437 第20回国体記念 20th National Athletic Meet
岐阜の初日用通信日付印(ハト印)を押印したカバーで、岐阜県警が国体時の交通規制について通知した郵便 Cover cancelled with the first-day comb type datestamp with dove mark. A letter from the Gifu Prefectural Police with notification of traffic regulations during the National Athletic Meet.

Type❷ FDC初日印局
FDC designated Post Offices (non-official)

FDC版元が初日印局として選定した局。大量の押印に対応するために態勢を整え、新切手が重点配備された。

P204 耶馬日田英彦山国定公園 Yaba-Hita-Hikosan Quasi-National Park
大分耶馬渓の櫛型印を押印した耶馬渓風物館差し出しのパノラマはがき Panoramic postcard cancelled with the Oita Yabakei comb type datestamp sent by the Yabakei Museum.

巻末資料

Type❸ 特印使用局
Post Offices using a Special Date Stamp

切手の地域性を考慮して当該切手について特印使用局として（追加）指定した局。特印定例局（1968年8月3日、郵政省告示第601号）のように、毎回特印を使用する局と切手の地域性に併せて使用する局がある。

c298 名古屋開府 350 年記念
350th Anniversary of Nagoya
千種の櫛型印を押印した名古屋大学教授差し出しのカバー
Cover cancelled with the Chigusa comb type datestamp sent by a professor at Nagoya University.

Type❹ 関連小型印使用局
Post Offices using a Commemorative Postmark

切手発行と同趣旨の小型印および切手発行を記念した小型印を使用した当該地域の局。

c308 天橋立
Amanohashidate (Bridge to Heaven)
発行記念の小型印を押印したカバー
Cover cancelled with a commemorative postmark for the stamp issue.

Type❺ 関連風景印使用局
Post Offices using a Related Pictorial Postmark

地域性を持つ切手発行に合わせて図柄を刷新もしくは使用開始した風景印使用局、および風景印にその図柄を描いている局（例：1961年花シリーズ）。

c335 花シリーズ・キキョウ Flower Series, Balloon Flower
初日印局沼田の風景印ときょうを描いた平湯局の風景印を押印したカバー
Cover cancelled with the pictorial postmark of Numata post office, designated as the first-day office, and the pictorial postmark of Hirayu post office with a balloon-flower design.

Type❻ 会場内局及び臨時出張所
Post Office at an Event Venue; Temporary Post Offices

スポーツ大会や国際会議・国際イベントに設けられた臨時出張所。会場内に存在する局または臨時出張所の担当局。

C418 東京 1964 オリンピック競技大会
Tokyo 1964 Olympic Games
フランスの選手がサインして差し出した東京オリンピック村の欧文三日月印を押印したカバー
Postcard signed by French Olympians and cancelled with the swordguard type datestamp of the Tokyo Olympic Village.

Type❼ 公園切手指定地域内局及び主要観光拠点局
Post Offices in National Park Stamp designated areas and Major Tourist Hubs

切手発行当時の指定地域内にあった局。指定地域内にはないが、主要な公園内観光地へのアクセス拠点となる隣接地域の局。

P91 富士箱根伊豆国立公園
Fuji-Hakone-Izu National Park
指定地域内局である箱根宮ノ下局の櫛型印を押印した非郵趣家カバー Non-philatelic cover cancelled with the comb type datestamp of Hakone-Miyanoshita, located in the designated "related area".

Type❽ 当該地域・施設の集配局
Collection and Delivery Post Offices for the Relevant Area/Facility

切手の発行趣旨にあたる特定地域または会場・施設などの集配を担当する局。

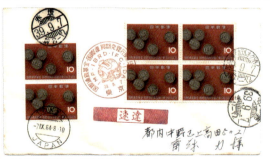

C413 国際通貨基金・国際復興開発銀行東京総会記念
IMF and IBRD Tokyo General Meeting
会場となり特印も使用された東京ホテルオークラ内局の消印（特印の局名表示は東京）も押印されたカバー Cover cancelled with several postmarks including that of Hotel Ohkura, Tokyo, where the meeting was held.

ご当地局（使用）一覧表 List of Related Area Post Office

★は「局名×消印」に希少性があるもの Scarce combinations are marked with a "★".

❶ 記念・特殊切手／国体、東京五輪を除く
Commemorative and Special Stamps / Excluding National Athletic Meet and Tokyo 1964 Olympic Games

番号 No.	切手名 Names	Type①、②	Type③	Type④	その他 Others
C225-226	UPU加入75年				
C227-228	日本赤十字社75年		芝・★日本赤十字社病院内（局名東京）		
C229	東京大学75年	本郷（局名東京）	目黒（局名東京）		
C232-235	立太子礼				
C236	電灯75年				
C237-238	皇太子帰朝				
C241	東京天文台75年	三鷹	★武蔵野、★境		
C242	男子スピードスケート	札幌			
C243	日本国際見本市	大坂中央、大阪東（局名表示大阪：3.20-4.23使用）	大阪港（局名大阪）		会場内に大阪中央、大阪東の臨時局あり⑥
C244	フリースタイルレスリング		代々木（局名東京：東京体育館）		
C247-248	ITU加盟75年				
CB1	趣味切手帳ペーン				
C249	国際商業会議所				
C252	趣味「ビードロ」				
C253	世界卓球		代々木（局名東京：東京都体育館内臨時局）⑥		
C254	世界柔道		浅草（局名東京：蔵前国技館内臨時局）⑥		
C255	世界こどもの日				
C256	東京開都500年			東京中央（記念郵便展）	
C257	佐久間ダム	静岡中部⑤			
C260	趣味「市川蝦蔵」				
C261	マナスル登頂				
C262	東海道電化	大津			
C263	日本機械巡航				
C264	国連加盟				
C265	国際地球観測年				
C266	原子炉完成	茨城東海			
C269	趣味「まりつき」				
C270	小河内ダム	小河内	★氷川、★古里、★沢井		
C271	製鉄100年	釜石	八幡、★八幡製鉄構内		
C272	関門トンネル	門司、下関		下関（記念切手展・下関大博覧会）、下関東・★長府（下関大博覧会）	
C273	趣味「雨中湯帰り」				
C274	日本開港100年	横浜、函館、長崎		横浜中（記念国際切手展）	
C275-278	アジア競技大会		代々木（局名東京：国立競技場）		
C279	ブラジル移住	横浜			
C280	国際胸部・器官食道				⑧晴海（東京第一生命ホール）、左京（京都大学）
C281	文通「京師」	京都中央（局名京都）			
C284	慶応義塾100年	慶応義塾前	★日吉		
C285	国際児童福祉		神田（局名東京：産経会館）		
C286	世界人権宣言				
C287	児島湾締切堤防	岡山	★岡山福島、倉敷、玉野、★児島		
C288	アジア文化会議				
C289-292	皇太子御成婚	宮内庁内			
C294	趣味「浮世源氏八景」				
C295	メートル法実施				
C296	赤十字思想				⑧王子（滝野川聖学院で記念式典）
C297	自然公園の日	日光、上九一色	★中禅寺・★日光湯本・★日光駅前（局名日光）		
C298	名古屋開府	名古屋中央（局名名古屋）	千種、名古屋東、名古屋中、昭和、中村、熱田、名古屋港（局名名古屋）	名古屋中央（名古屋城再建、記念切手展）、千種、名古屋東、名古屋中、昭和、熱田、名古屋港（名古屋城再建：局名名古屋）	
C299	文通「桑名」	桑名			
C300	国際航空運送				⑥帝国ホテル内
C303	松蔭100年祭	萩、宇部		萩（松陰百年祭）	
C304	GATT東京総会		神田（局名東京：産経会館）		
C305	尾崎記念会館	国会内（局名東京）			
C306	奈良遷都1250年	奈良	★大和高田、★橿原、★大和郡山、★五条、★天理、★下市、★桜井		
C307	日本三景 松島	松島海岸		松島海岸、松島、塩釜、仙台（発ации記念）、★仙台（記念切手展）	
C308	日本三景 天橋立	天橋立		天橋立、天橋立駅前、宮津、岩滝（発行記念）	
C309	日本三景 宮島	宮島		宮島（発行記念）	⑦宮島口
C310	趣味「伊勢」				

157

巻末資料

番号	切手名	Type①、②	Type③	Type④	その他
C311-313	日米修好通商	国会内 (局名東京)			
C314	鳥類保護会議		麻布 (局名東京：麻布国際文化会館)、相川、★両津、★羽咋、★輪島		
C315	国際放送25年		★東京都麹町内幸町 (局名東京)		
C316	ハワイ官約移住	横浜			
C317	航空50年	代々木 (局名東京)			
C318-319	列国議会同盟	国会内 (局名東京)			
C320	文通「蒲原」	蒲原			
C321	岡山天体観測所	鴨方	矢掛、金光、★武蔵野、★調布、★三鷹大沢		
C324	白瀬中尉南極	金浦	本荘		
C325-326	議会開設70年	国会内 (局名東京)			
C327	花 スイセン	四箇浦⑤		福井 (花切手展)	★⑤越廼 (四箇浦と同一風景印)
C328	花 ウメ	水戸、太宰府			⑤青梅、熱海、月ケ瀬
C329	花 ツバキ	大島			⑤波浮港、泉津、新島、椿
C330	花 ヤマザクラ	吉野		吉野 (発行記念)	⑤菊池
C331	花 ボタン	大根島		大根島 (切手展)、松江 (切手展)	⑤須賀川、初瀬、当麻、東松山
C332	花 ハナショウブ	葛飾			⑤知立
C333	花 ヤマユリ	横須賀			⑤斜里
C334	花 アサガオ	下谷入谷⑤		松任 (切手展)	★⑤下谷坂本町 (下谷入谷と同一風景印)
C335	花 キキョウ	沼田			
C336	花 リンドウ	坊中⑤			⑤天ケ瀬
C337	花 キク	笠間			
C338	花 サザンカ	仁比山⑤			
C339	郵便90年				
C340	趣味「舞妓図」				
C341	国際ロータリー		★帝国ホテル内 (局名：東京 ※欧文特印も使用)		
C342	愛知用水	名古屋中央 (局名名古屋)	千種、名古屋東、名古屋中、熱田		
C343	日本標準時	明石	大久保		
C346	文通「箱根」	箱根町			
C373	国立国会図書館	国会内 (局名東京)			
C374	行事 ひなまつり	中京		中京、京都中央 (ひな人形展：局名京都)	
C375	行事 たなばた	仙台⑤		仙台 (切手展)、★島根大東 (切手展)	⑤平塚
C376	行事 七五三	渋谷			
C377	行事 節分	成田			
C378	趣味「花下遊楽図」				
C379	北陸トンネル	敦賀、今庄		福井 (のりもの切手展)	
C380	アジアジャンボリー	御殿場			
C381	若戸大橋	若松⑤、戸畑⑤			
C382	文通「日本橋」	日本橋			
C385	北九州市発足	若松、戸畑、小倉、八幡、門司 (局名北九州)			
C386	飢餓救済				
C387	趣味「千姫」				
C388	赤十字規約制定		新宿 (局名東京：厚生年金会館)		
C389	国際かんがい排水				
C390	鳥 ルリカケス	名瀬		★名瀬 (奄美大島日本復帰10年)	
C391	鳥 ライチョウ	富山、松本、★上高地		富山 (切手展)、松本 (切手展)	
C392	鳥 キジバト	越谷			
C393	鳥 コウノトリ	豊岡		大阪中央 (切手展)	⑤豊岡駅前
C394	鳥 ウグイス	牧岡		牧岡 (切手展)	
C395	鳥 ホオジロ	佐倉			
C396	名神高速道路	京都中央 (局名京都)	伏見、中京、大津、草津、石部		
C397	ガールスカウト	戸隠			
C398	国際電波科学連合	高輪 (局名東京)			
C399	文通「神奈川沖浪裏」	神奈川			
C400	東京国際スポーツ	代々木 (局名東京：国立競技場に臨時局)⑥、新宿			
C403	祭り 高山祭	高山		高山 (切手展)	
C404	祭り 祇園祭	祇園			
C405	祭り 相馬野馬追	原町、相馬			
C406	祭り 秩父夜祭	秩父		秩父 (切手展)	
C407	趣味「宿木」				
C408	姫路城修理	姫路	★姫路駅前 (局名姫路)		
C411	太平洋横断ケーブル	二宮			
C412	首都高速道路	日本橋⑤ (局名東京)			
C413	国際通貨基金	日本橋 (局名東京)	赤坂・ホテルオークラ内 (局名東京)		
C420	八郎潟干陸式	船越⑤、天王	秋田・秋田県庁構内 (局名秋田)、(以下★) 八龍、鹿渡、一日市、八郎潟、大久保、琴浜		
C421	東海道新幹線	東京中央 (局名東京)、大阪中央 (局名大阪)	横浜、小田原、熱海、静岡、浜松、豊橋、名古屋中央、岐阜羽島、滋賀米原、京都中央	大阪中央 (切手展)	
C422	文通「程ヶ谷」	保土ヶ谷			

番号	切手名	Type①、②	Type③	Type④	その他
C423	富士山頂気象レーダー	富士宮	御殿場、★気象庁内		
C424	通信総合博物館	東京中央（局名東京）⑥			
C425	趣味「序の舞」				
C426	国立こどもの国	長津田	★港北		
C427	国土緑化運動	大山、米子			
C428	ITU100年	高輪（局名東京）			
C429	国際協力年				
C430	海の記念日	横浜	★名古屋港		
C431	愛の血液助け合い				
C532	国際原子力機関				初日ハト印開始
C433	国勢調査				
C434	文通「三坂水面」	富士吉田			
C435	国民参政75周年				国会内⑧
C438	国際耳鼻咽喉・小児科				
C439	南極地域観測		ふじ船内、昭和基地内		
C440	電話創業75年		麹町、下谷、横浜中		
C441	魚介 イセエビ	牛深			
C442	魚介 コイ	野沢	野沢・中込・★野沢南（佐久鯉まつり）、蟹江（切手展）		
C443	魚介 マダイ	鞆、小湊	鞆・★福山		
C444	魚介 カツオ	焼津、土佐清水	焼津（切手展）、土佐清水（切手展）		
C445	魚介 アユ	郡上八幡⑤	郡上八幡（切手展）		
C453	名園「偕楽園」	水戸	水戸（観梅70周年）		
C456	国際工業所有権				
C457	趣味「蝶」				

❷ 国民体育大会、東京1964オリンピック競技大会 National Athletic Meet and Tokyo 1964 Olympic Games

番号 No.	切手名 Name	Type①、②	Type③	Type④	その他
C230-231	第7回国体	仙台、福島、山形	★長町、会津若松、郡山、平、白河、飯坂、喜多方、★福島東山、★福島湯野、★福島山郷、米沢、鶴岡、酒田、新庄、★山形吹浦	仙台、福島、山形、★山形吹浦、会津若松、郡山、白河、平、飯坂、★福島湯野、喜多方、酒田、★福島山郷（競技別）	
C239-240	第8回国体	松山、高松、徳島、高知	道後・三津浜（局名松山）、今治、西条、新居浜、八幡浜、宇和島、観音寺、丸亀、坂田、善通寺、多度津、★高松五番丁・高松牟礼・高松屋島・高松長尾・高松内海、高砂、西脇、洲本、宝塚、鳴門、小松島、★阿波池田	松山、道後・三津浜（局名松山）、新居浜、高松、高松五番丁（局名高松）、徳島、鳴門、高知、丸亀、西条、善通寺、小松島、坂出、今治、八幡浜、多度津（競技別）	
C245-246	第9回国体	旭川、小樽、札幌	函館、室蘭、東室蘭、釧路、帯広、岩見沢、滝川、稚内、網走、北見、根室、石狩深川、夕張、美唄、岩内、名寄、倶知安、余市、留萌、苫小牧、江差、寿都、浦河、砂川、八雲、江別、美幌、士別、十勝池田、富良野、栗山、伊達、幾春別、我路、赤平、胆振千歳、紋別、遠軽、中標津（但し同一市内は局名市名）★小型印使用局以外は★	旭川、小樽、札幌、函館、美唄、岩見沢、苫小牧、札幌（記念スポーツ切手展）	
C250-251	第10回国体	横浜	鶴見、神奈川、横浜中、磯子、保土ヶ谷、戸塚、横浜桜木、中原、川崎、溝ノ口、横須賀、久里浜、浦田、逗子、平塚、鎌倉、大船、藤沢、茅ヶ崎、厚木、小田原、湯河原、箱根湯本、秦野 ★小型印使用局以外は★	神奈川、綱島、横浜、茅ヶ崎、★茅ヶ崎茶屋町、★茅ヶ崎海岸、鎌倉、★鎌倉雪ノ下、鶴見、久里浜、保土ヶ谷、浦田、追浜、川崎、藤沢、藤倉、小田原、平塚、秦野、磯子、★横浜富岡、伊勢原、綱島	
C258-259	第11回国体	西宮、神戸中央	長田、灘、須磨、垂水、御影、姫路、飾磨、広畑、網干、尼崎、芦屋、明石、加古川、高砂、西脇、洲本、宝塚、大久保、龍野、森	西宮、神戸中央、灘、高砂、尼崎、明石、宝塚、姫路、赤穂、加古川、西脇、長田、高砂、洲本、芦屋、龍野、洲本（競技別）	
C267-268	第12回国体	静岡	浜松、沼津、清水、熱海、三島、吉原、富士、磐田、藤枝、島田、掛川、小山、蒲原、大仁、井川	静岡、浜松、沼津、掛川、小山、熱海、清水、三島、富士、島田、磐田、吉原、大仁、蒲原、井川、藤枝（競技別）	
C282-283	第13回国体	富山	高岡、魚津、伏木、氷見、新湊、岩瀬、滑川、石動、礪波、越中八尾、黒部、立山、泊、上滝、福野、井波、笹津、上市、庄川、福光、婦中	富山、高岡、滑川、笹津、婦中、石動、上市、岩瀬、氷見、立山、庄川、礪波、井波、福野、黒部、新湊、泊、★伏木、魚津、★越中八尾、★福光、★上滝（競技別）	
C301-302	第14回国体	東京中央	神田・麹町・日本橋・京橋・下谷・新宿・代々木・牛込・小石川・玉川・品川・本郷・蒲田・浅草・中野・深川・足立・赤坂・麻布・本所・芝・千歳・渋谷（局名：）★立川、武蔵府中、★八王子、★奥多摩、★東村山、★蕨		
C322-323	第15回国体	熊本	坪井（局名熊本）	熊本・坪井（熊本）、八代・日奈久（八代）、坪井、水俣、人吉、荒尾、玉名、山鹿、菊池、★本渡、★松橋、★御船、★北部、★宇土、★河蘇、★坊中・本妙（阿蘇）、赤水（阿蘇）、鹿児島、鹿児島東（鹿児島）	
C344-345	第16回国体	秋田	秋田駅前、秋田県庁構内、能代、横手、大館、土崎、大曲、湯沢、本荘	能代（体操競技）、秋田、土崎、大館、横手、大曲、湯沢、本荘、（以下）★鷹巣、新屋、六郷、象潟、由利院内、五城目、男鹿、大久保、二ツ井、平沢、角館、生保内、八幡平（大会記念）	
C383-384	第17回国体	岡山	岡山南、倉敷、津山、玉野、児島、玉島、笠岡、西大寺、井原、総社、高梁、新見、備前、美作勝山、金光、真金、湯原、久世、野谷（すべて★）		
C401-402	第18回国体	長門、山口	萩、小郡、長府、防府、岩国、宇部、大田、光、小野田、柳井、長門、田布施、安岡、下関東、徳山、厚狭、下松、秋吉（すべて★）	長門、山口（大会記念）、下関（切手展）	
C409-410	第19回国体	新潟	長岡、高田、三条、新発田、柏崎、直江津、両津、新津、加茂、村上、★水原、巻、★松ケ崎、★大野町、弥彦、柿崎、鹿瀬、黒川、★下関	新潟、長岡（大会記念）	
C436-437	第20回国体	岐阜	大垣、高山、多治見、関、中津川、笠松、美濃加茂、各務原、郡上八幡、土岐、瑞浪、美濃、羽島、★垂井、揖斐川、穂積、富富	岐阜、大垣、高山、多治見、中津川、美濃加茂、各務原、土岐、羽島、穂積（大会記念）、岐阜（切手展）	
C414-418	東京1964大会	（C414のみ）鹿児島、宮崎、札幌中央	東京五輪村⑥、赤坂・玉川・牛込・渋谷・本郷・麹町・小石川（局名東京）、神奈川・横浜中（局名横浜）、大宮、千葉、葉山、★二宮、★所沢・八王子・藤沢・相模湖・蕨・大和・軽井沢	東京五輪村⑥、赤坂、玉川、神奈川、大宮、牛込、渋谷、千葉、新宿、横浜大、葉山、二宮、麹町、代々木、小石川、八王子、藤沢、相模湖、蕨、大和、軽井沢	

❸ 第2次国立公園切手 2nd National Park Series

番号 No.	切手名 Names	Type①、②	Type④	Type⑤	Type⑦
P91-94	富士箱根伊豆	箱根町、南崎、船津、三津	伊東「発行記念切手とスタンプ展」	三津	多数あり
P95-98	日光	片品、那須、中禅寺、塩原			日光
P99-100	雲仙天草	雲仙、松島			
P101-102	白山	白峰	金沢「発行記念切手展」		
P103-104	磐梯朝日	大井沢、裏磐梯	山形、福島「発行記念切手展」		西五百川、下関
P105-106	瀬戸内海	鳴門、児島	徳島「発行記念観光展」、岡山「発行記念世界の切手展」、児島「発行記念」		下津井、高松、明石
P107-108	大雪山	鹿追、層雲峡	旭川「指定30年」	鹿追、★瓜幕	上川、中士幌
P109-110	伊勢志摩	伊勢、鳥羽			二見、大王、五十鈴川
P111-112	大山隠岐	大山、米子、西郷		西郷	岸本
P113-114	上信越高原	越後田沢、野尻湖	信濃町、妙高「妙高高原野尻湖観光祭」	越後田沢	中郷、上田
P115-116	阿蘇	宮地、阿蘇		赤水	熊本、小国、坊中
P117-118	知床	斜里、羅臼			

❹ 国定公園切手 Quasi-National Park Series

番号 No.	切手名 Names	Type①、②	Type④	Type⑤	Type⑦
P200-201	佐渡弥彦	相川、姫津、小田、弥彦		相川、姫津、小田、弥彦	河原田、両津、新潟
P202-203	秋吉台	秋吉	★山口「発行記念切手展」	秋吉	
P204-205	耶馬日田英彦山	耶馬溪、日田、彦山	★日田「第3集発行記念切手展」	彦山	山国、南院内
P206	三河湾	蒲郡			豊橋
P207	網走	浜小清水	網走「発行記念切手展」、浜小清水「観光センター完成」		北見呼人
P208	足摺	土佐清水	土佐清水「指定5周年記念」		
P209	南房総	白浜		白浜、小戸	館山
P210	琵琶湖	石山	石山、大津、★彦根、★近江八幡、★長浜、★草津「発行記念」		
P211	山陰海岸	鳥取	鳥取「発行記念」		
P212	大沼	大沼			函館
P213	北長門海岸	長門	長門「発行記念」	長門	萩
P214	錦江湾	鹿児島			南桜島、東桜島
P215	金剛生駒	千早	阿倍野「発行記念写真展」	千早	大和郡山、生駒
P216	水郷	潮来、佐原		潮来	
P217	石鎚	石鎚	松山「観光展」	石鎚	小松
P218	玄海	芥屋			呼子、唐津
P219	伊豆七島	八丈島		★末吉	伊豆七島各島
P220	若狭湾	高浜		高浜	敦賀
P221	日南海岸	宮崎		宮崎	
P222	ニセコ積丹小樽	昆布			小樽、昆布温泉
P223	蔵王	遠刈田			
P224	室戸阿南	室戸、日和佐		日和佐	室戸岬

❺ 年賀切手 New Year's Greeting Stamps

番号 No.	切手名 Names	Type①、②	Type④	Type⑤
N7	27年 翁の面			
N8	28年 三番叟			
N9	29年 三春駒	三春		
N10	30年 加賀起き上がり	金沢	金沢（発行記念切手展）	
N11	31年 こけし	白石	白石（発行記念こけし祭）	
N12	32年 だんじり	長崎		長崎
N13	33年 犬はりこ	浅草		浅草
N14	34年 鯛えびす	高松	高松（郷土人形展）	
N15	35年 米食いネズミ	金沢	金沢（郷土玩具と切手展）	
N16	36年 赤べこ	会津若松、盛岡	会津若松（赤べこまつり12/20、発行記念SS1/20）、盛岡（郷土玩具展）	
N17	37年 張り子のトラ	出雲	出雲（発行記念切手玩具展）	
N18	38年 ウサギ	鹿島（佐賀県）	鹿島（発行記念切手展）	
N19	39年 辰と竜神	龍王（山梨県）、岩井（鳥取県）	岩井（発行記念）、甲府（玩具と切手展）	龍王
N20	40年 麦わらヘビ	浅草	浅草（発行s切手展）	
N21	41年 しのび駒	花巻		

郵便料金変遷表 Transition of Postage Rates

国内郵便 Domestic Mail

❶ 国内通常郵便　Domestic Regular Mail

年月日 Date	書状 Envelope		はがき Postal Card			定期刊行物 Periodical		印刷物 Printed Matter			農産物種子 Agriccultual Seeds
	基本 basic	簡易書簡 kan'i shokan	普通はがき ordinary	往復はがき double reply	小包はがき parcel postal card	基本 basic	月3回以上 at least 3 times a month	基本 basic	通信教育 correspondence education	点字 braille	
	20g毎 every 20g	（売価） (selling price) (年賀状使用) (new year's)				100g毎 every 100g	100g毎 every 100g	100g毎 every 100g	100g毎 every 100g	1kg毎 every 1kg	100g毎 every 100g
1951(昭和26).11.1	10 yen	10 yen (11 yen)	5 yen (4 yen)	10 yen	6 yen	4 yen	1 yen	8 yen	4 yen	1 yen	2 yen
1958(昭和33).11.20		10 yen (12 yen)									
1961(昭和36).6.1						6 yen	2 yen	50g毎 every 50g 10 yen		1kg迄 up to 1kg 無料 free	
1966(昭和41).7.1	25g迄 up to 25g 15 yen	郵便書簡 letter sheet 15 yen	7 yen	14 yen	8 yen	100g迄 up to 100g 3 yen		学術 academic 100g毎 every 100g 10 yen			6 yen

❷ 国内航空郵便　Domestic Airmail

年月日 Date	書状 Envelope	はがき Postal Card	印刷物 Printed Matter
	20g迄 up to 20g		20g迄 up to 20g
1951(昭和26).11.1	25 yen	15 yen	20 yen

※1953(昭和28).7.14 速達と統合 Integrated into Special Derivery

❸ 国内特殊取扱郵便　Domestic Special Treatments

年月日 Date	書留 Registered			速達 Special Delivery		配達証明 Certification of Delivery		引受時刻証明 Certification of Acceptance Time		代金引換 Cash on Delivery	訴訟書類 Court Docu-ments
	基本料金 basic charge	加算料金 additional surcharge		書状・はがき Envelope and Postal Card	小包 Postal Parcel	差出時請求 on delivery	差出後請求 after delivery	引受時刻 accept -ance time	書留書状 Registered Envelope		
		現金 for cash	物品 for goods								
1951(昭和26).11.1	1,000円迄 up to 1,000 yen 35 yen	1,000円超2,000円毎 every 2,000 yen over 1,000 yen 5 yen	1 yen	25 yen	40 yen	50 yen	75 yen	50 yen	45 yen	50 yen	50 yen
1953(昭和28).7.5					50 yen						
1961(昭和36).6.1	1,000円迄 up to 1,000 yen 40 yen			30 yen	70 yen	60 yen	90 yen	60 yen	50 yen	60 yen	80 yen
1966(昭和41).7.1	3,000円迄 up to 3,000 yen 60 yen	3,000円超2,000円毎 every 2,000 yen over 3,000 yen 5 yen	1 yen	200g迄 up to 200g 50 yen	80 yen	70 yen	120 yen	70 yen	75 yen	80 yen	100 yen

❹ 国内小包郵便 Domestic Postal Parcel

年月日 Date	重量 Weight	第1地帯 1st zone		第2地帯 2nd zone	第3地帯 3rd zone
		市内 city	市外 outside city		
1951（昭和26）.11.1	2kg迄 up to 2kg	30 yen	50 yen	65 yen	85 yen
	2kg超4kg迄 up to 4kg over 2kg	45 yen	65 yen	80 yen	100 yen
	4kg超6kg迄 up to 6kg over 4kg	60 yen	80 yen	95 yen	115 yen
1953（昭和28）.7.5	2kg迄 up to 2kg	30 yen	55 yen	70 yen	90 yen
	2kg超4kg迄 up to 4kg over 2kg	45 yen	75 yen	95 yen	120 yen
	4kg超6kg迄 up to 6kg over 4kg	60 yen	95 yen	120 yen	150 yen
1961（昭和36）.6.1	2kg迄 up to 2kg	50 yen	90 yen	120 yen	170 yen
	2kg超2kg毎 every 2kg over 2kg	20 yen	30 yen	40 yen	60 yen
1966（昭和41）.4.1	2kg迄 up to 2kg	70 yen	120 yen	160 yen	230 yen
	2kg超2kg毎 every 2kg over 2kg	20 yen	30 yen	40 yen	50 yen

外国郵便 International Mail

❺ 1950（昭和25）.9.1～1966（昭和41）.6.30　平面路（船便）通常郵便　International Ordinary Sea Mail

年月日 Date	書状 Envelope		はがき Postal Card	印刷物 Printed Matter		点字 Braille	小形包装物 Small Packet		業務用書類 Business Documents		商品見本 Sample		書留 Registered	別配達 Special Delivery
	20g迄 up to 20g	20g超20g毎 every 20g over 20g		50g迄 up to 50g	50g超50g毎 every 50g over 50g	1kg毎 every 1kg	250g迄 up to 250g	200g超50g毎 every 50g over 200g	200g迄 up to 200g	200g超50g毎 every 50g over 200g	50g迄 up to 50g	50g超50g毎 every 50g over 50g		
1950（昭和25）.9.1	24 yen	14 yen	14 yen	5 yen	5 yen	2 yen	48 yen	10 yen	24 yen	5 yen	10 yen	5 yen	48 yen	72 yen
1951（昭和26）.12.1				10 yen	5 yen									
1953（昭和28）.7.1						1 yen								
1959（昭和34）.4.1	30 yen	20 yen	20 yen				60 yen	10 yen	30 yen	5 yen			50 yen	70 yen
1961（昭和36）.10.1	40 yen	25 yen	25 yen	15 yen	10 yen	無料 free	80 yen	15 yen	40 yen	10 yen	15 yen	10 yen		
1966（昭和41）.7.1	50 yen	30 yen	30 yen	25 yen	10 yen		100 yen	20 yen			25 yen		70 yen	95 yen

❻-1　1949（昭和24）.6.1～1951（昭和26）.11.30　航空通常郵便（航空料金割増時代）International Air Mail（period of Air Surcharge）

年月日 Date	書状・はがき Envelope and Postal Card							
	基本料金 basic fee			割増し料金（10g毎）additional fees (every 10 g)				
	書状 Envelope		はがき Postal Card	第1地帯 1st zone	第2地帯 2nd zone	第3地帯 3rd zone	第4地帯 4th zone	第5地帯 5th zone
	20g迄 up to 20g	20g超20g毎 every 20g over 20g		10g毎 every 10g				
1949（昭和24）.6.1	24 yen	14 yen	14 yen	16 yen	34 yen	59 yen	103 yen	144 yen

年月日 Date	印刷物 Printed Matter							
	基本料金 basic fees			割増料金 additional fees				
	50g迄 up to 50g	50g超50g毎 every 50g over 50g		第1地帯 1st zone	第2地帯 2nd zone	第3地帯 3rd zone	第4地帯 4th zone	第5地帯 5th zone
				10g毎 every 10g				
1951（昭和26）.4.1	5 yen	5 yen		7 yen	12 yen	18 yen	36 yen	41 yen

❻-2　1951（昭和26）.12.1.～1966（昭和41）.6.30　航空通常郵便（航空料金合算時代）International Air Mail（Piriod of Combined Fee）

年月日 Date	書状 Envelope					はがき Postal Card				
	第1地帯 1st zone	第2地帯 2nd zone	第3地帯 3rd zone	第4地帯 4th zone	第5地帯 5th zone	第1地帯 1st zone	第2地帯 2nd zone	第3地帯 3rd zone	第4地帯 4th zone	第5地帯 5th zone
	10g毎 every 10g									
1951（昭和26）.12.1	40 yen	55 yen	80 yen	125 yen	160 yen	40 yen	55 yen	80 yen	125 yen	160 yen
1953（昭和28）.7.1	35 yen	50 yen	70 yen	115 yen	145 yen	30 yen	35 yen	40 yen	60 yen	75 yen
1959（昭和34）.4.1	40 yen	75 yen	115 yen			35 yen	40 yen	60 yen		
1961（昭和36）.10.1	50 yen	80 yen	100 yen			30 yen		50 yen		
1966（昭和41）.7.1	60 yen	90 yen	110 yen			35 yen	45 yen	55 yen		

次ページにつづく Continued to next page

年月日 Date	印刷物・点字 Printed Matter and Braille					小形包装物 Small Packet				
	第1地帯 1st zone	第2地帯 2nd zone	第3地帯 3rd zone	第4地帯 4th zone	第5地帯 5th zone	第1地帯 1st zone	第2地帯 2nd zone	第3地帯 3rd zone	第4地帯 4th zone	第5地帯 5th zone
	20g毎 every 20g					上段：60g迄、下段：60g超20g毎 upper: up to 20g lower: every 20g over 60g				
1951 (昭和26).12.1	25 yen	30 yen	40 yen	75 yen	85 yen	25 yen / 25 yen	30 yen / 30 yen	40 yen / 40 yen	75 yen / 75 yen	85 yen / 85 yen
1953 (昭和28).7.1			35 yen	70 yen	80 yen	75 yen / 25 yen	90 yen / 30 yen	105 yen / 35 yen	210 yen / 70 yen	240 yen / 80 yen
1959 (昭和34).4.1			60 yen			75 yen / 25 yen	90 yen / 30 yen	180 yen / 60 yen		
1961 (昭和36).10.1	30 yen	40 yen	50 yen			90 yen / 30 yen	120 yen / 30 yen	150 yen / 30 yen		
1966 (昭和41).7.1	上段：20g迄、下段：20g超20g毎 upper: up to 20g lower: every 20g over 20g					上段：80g迄、下段：80g超20g毎 upper: up to 80g lower: every 20g over 80g				
	40 yen / 30 yen	50 yen / 40 yen	60 yen / 40 yen			130 yen / 30 yen	170 yen / 40 yen	210 yen / 50 yen		

年月日 Date	業務用書類 Business Documents			商品見本 Sample		
	第1地帯 1st zone	第2地帯 2nd zone	第3地帯 3rd zone	第1地帯 1st zone	第2地帯 2nd zone	第3地帯 3rd zone
	上段：60g迄、下段：60g超20g毎 upper: up to 20g lower: every 20g over 60g			上段：60g迄、下段：60g超20g毎 upper: up to 20g lower: every 20g over 60g		
1959 (昭和34).4.1	35 yen / 25 yen	40 yen / 30 yen	65 yen / 60 yen	25 yen / 25 yen	30 yen / 30 yen	60 yen / 60 yen
1961 (昭和36).10.1	45 yen / 30 yen	50 yen / 40 yen	55 yen / 50 yen	30 yen / 30 yen	40 yen / 40 yen	50 yen / 50 yen
1966 (昭和41).7.1	廃止 abolition			上段：60g迄、下段：60g超20g毎 upper: up to 20g lower: every 20g over 60g		
				40 yen / 30 yen	50 yen / 40 yen	60 yen / 50 yen

❻-3 1949 (昭和24).3.1～1966 (昭和41).6.30 航空書簡 Aerogramme

年月日 Date	航空書簡 Aerogramme
1949 (昭26).6.1	62yen
1952 (昭和27).4.1	50yen
1953 (昭和28).7.1	45yen
1961 (昭和36).10.1	50yen

❼ 1949 (昭和24).6.1.～1966 (昭和41).6.30　地帯別外国郵便 International Mail by Zone

年月日 Date	第1地帯 1st zone	第2地帯 2nd zone	第3地帯 3rd zone	第4地帯 4th zone	第5地帯 5th zone
1949 (昭和24).6.1	琉球、朝鮮、中華民国、香港、マカオ Ryukyus, Korea, China, Hong Kong and Macau	ウェーキ、フィリピン、シャム、グアム、アラスカ Wake Island, Filipinas, Siam, Guam abd Alaska	パキスタン、インド、ハワイ、アメリカ、カナダ Pakistan, India, Hawaii, U.S.A. and Canada	マラヤ、セイロン、ビルマ、オーストラリア、近東、ヨーロッパ、中米 Malaya, Ceylon, Burma, Australia, Near East, Europe and Central America	ニュージーランド、南米、アフリカ New Zealand, South America and Africa
1951 (昭和26).12.1	中国、台湾、香港、マカオ China, Taiwan, Hong Kong and Macau	フィリピン、タイ、グアム、アラスカ、南洋諸島 Filipinas, Thailand, Guam, Alaska and South Sea Islands	ハワイ、アメリカ、カナダ、オーストラリア Hawaii, U.S.A., Canada and Australia	オセアニア、近東、ヨーロッパ、中米、北アフリカ Oceania, Near East, Europe, Central America and North Africa	その他アフリカ Africa outside North Africa
1953 (昭和28).7.1	朝鮮、中国、台湾、香港、マカオ Korea, China, Taiwan, Hong Kong and Macau	ビルマ、マラヤ、フィリピン、タイ、グアム、アラスカ Burma, Malaya, Philipines, Thailand, Guam and Alaska	その他アジア、ハワイ、アメリカ、カナダ、オーストラリア Other Asia, Hawaii, U.S.A., Canada and Australia	ニュージーランド、近東、中南米、北アフリカ、ソ連 New Zealand, Near East, Central America, South America, North Africa and U.S.S.R.	その他アフリカ Africa outside North Africa
1959 (昭和34).4.1	東アジア、オセアニア East Asia and Oceania	アメリカ、カナダ、中米、西インド諸島 U.S.A., Cabada, Central America and West Indies	中近東、ヨーロッパ、南米、ソ連、アフリカ Middle and Near East, Europe, South America, U.S.S.R. and Africa		

西暦対照表 Comparative List of Japanese, Chinese, Manchuria and Gregorian Calenders

西暦 Gregorian	和暦 Japanese	中国暦 Chinese	満州暦 Manchuria
1860	安政 Ansei 7	咸豊 Kanpo 10	
1861	万延 Man'en 2		
1862	文久 Bunkyu 2	同治 Douji 1	
1863	3	2	
1864	4	3	
1865	元治 Genji 2	4	
1866	慶応 Keio 2	5	
1867	3	6	
1868	4	7	
1869	明治 Meiji 2	8	
1870	3	9	
1871	4	10	
1872	5	11	
1873	6	12	
1874	7	13	
1875	8	光緒 kosho 1	
1876	9	2	
1877	10	3	
1878	11	4	
1879	12	5	
1880	13	6	
1881	14	7	
1882	15	8	
1883	16	9	
1884	17	10	
1885	18	11	
1886	19	12	
1887	20	13	
1888	21	14	
1889	22	15	
1890	23	16	
1891	24	17	
1892	25	18	
1893	26	19	
1894	27	20	
1895	28	21	
1896	29	22	
1897	30	23	
1898	31	24	
1899	32	25	
1900	33	26	
1901	34	27	
1902	35	28	
1903	明治 Meiji 36	光緒 kosho 29	
1904	37	30	
1905	38	31	
1906	39	32	
1907	40	33	
1908	41	34	
1909	42	宣統 Sento 1	
1910	43	2	
1911	44	3	
1912	45		
1913	大正 Taisho 2		
1914	3		
1915	4		
1916	5		
1917	6		
1918	7		
1919	8		
1920	9		
1921	10		
1922	11		
1923	12		
1924	13		
1925	14		
1926	15		
1927	昭和 Showa 2		
1928	3		
1929	4		
1930	5		
1931	6		
1932	7		大同 Daido 1
1933	8		2
1934	9		康徳 Kotoku 1
1935	10		2
1936	11		3
1937	12		4
1938	13		5
1939	14		6
1940	15		7
1941	16		8
1942	17		9
1943	18		10
1944	19		11
1945	20		12
1946	昭和 Showa 21		
1947	22		
1948	23		
1949	24		
1950	25		
1951	26		
1952	27		
1953	28		
1954	29		
1955	30		
1956	31		
1957	32		
1958	33		
1959	34		
1960	35		
1961	36		
1962	37		
1963	38		
1964	39		
1965	40		
1966	41		
1967	42		
1968	43		
1969	44		
1970	45		
1971	46		
1972	47		
1973	48		
1974	49		
1975	50		
1976	51		
1977	52		
1978	53		
1979	54		
1980	55		
1981	56		
1982	57		
1983	58		
1984	59		
1985	60		
1986	61		
1987	62		
1988	63		
1989	昭和 Showa 64		
1990	平成 Heisei 2		
1991	3		
1992	4		
1993	5		
1994	6		
1995	7		
1996	8		
1997	9		
1998	10		
1999	11		
2000	12		
2001	13		
2002	14		
2003	15		
2004	16		
2005	17		
2006	18		
2007	19		
2008	20		
2009	21		
2010	22		
2011	23		
2012	24		
2013	25		
2014	26		
2015	27		
2016	28		
2017	29		
2018	30		
2019	31		
2020	令和 Reiwa 2		
2021	3		
2022	4		
2023	5		
2024	6		
2025	(7)		
2026	(8)		
2027	(9)		

1) 和暦は、明治5年まで旧暦（太陰暦）を採用。
2) 日本は1872年（明治5）11月9日に旧暦（大陰暦）を廃止して新暦（太陽暦）を採用、旧暦の明治5年12月3日は新暦の明治6年1月1日。
3) 1867年8月に明治天皇即位、1868年（慶応4）9月8日から明治元年。
4) 1912年（明治45）7月30日に明治天皇崩御、同日付で大正に改元。郵便局の日付印は、1931年7月31日から大正元年。
5) 1926年（大正15）12月25日に大正天皇崩御、同日付で昭和に改元。郵便局の日付印は、1926年12月25日から昭和元年（一部の局のみ）。
6) 1989年（昭和64）1月7日に昭和天皇崩御、1月8日から平成に改元。郵便局の日付印は、1989年1月8日から平成元年。
7) 2019年（平成31）4月30日に新天皇即位、5月1日から令和に改元。郵便局の日付印は、2019年5月1日から令和元年。

1) The Japanese calendar is the lunar calendar until Meiji 5.
2) In Japanese calendar, 1872 (Meiji 5) .12.3 of the lunar calendar is 1868 (Meiji 6) .1.1 of the solar calendar.
3) Emperor Meiji was crowned in August 1867, from 1868 (Keio 4) .9.8 to 1st year of Meiji.
4) Emperor Meiji died on 1912 (Meiji 45) .7.30 and was converted to Taisho on the same day. The date stamp of the post office is from 1912.7.31 to 1st year of Taisho.
5) Emperor Taisho died on 1926 (Taisho 15) .12.25 and was converted to Showa on the same day. The date stamp of the post office is from 1926.12.25 to 1st year of Showa.
6) Emperor Showa died on 1989 (Showa 64) .1.7 and was converted to Heisei from 1989.1.8. The date stamp of the post office is from 1989.1.8 to 1st year of Heisei.
7) The New Emperor was crowned on 2019 (Heisei 31) .4.30 and was converted to Reiwa from 2019.5.1. The date stamp of the post office is from 2019.5.1 to 1st year of Reiwa.

企画・監修 Planning and General Editing

公益財団法人日本郵趣協会 専門カタログ・ワーキンググループ 座長：稲葉良一 委員：石川勝己、那須伊允、山口 充、山田廉一、横山裕三	Specialized Catalogue Working Group of Japan Philatelic Society Chairman：Inaba Ryoichi Member：Ishikawa Katsumi, Nasu Tadanobu, Yamaguchi Mitsuru, Yamada Renichi, Yokoyama Hiromi

執筆・図版 Author

大久保幸夫、永吉秀夫、山田廉一、横山裕三、 ケネス・J・ブライソン	Okubo Yukio, Nagayoshi Hideo, Yamada Renichi, Yokoyama Hiromi, Kenneth J. Bryson

資料協力 Supporter

濱谷彰彦、記念特殊切手研究会、切手の博物館、 郵趣サービス社	Hamaya Akihiko, Commemorative and Special Stamp Study Group, Mizuhara Memorial Philatelic Museum, Japan Philatelic Co.,Ltd.

参考文献 References

―書籍 Philatelic Books―
『日本切手の製造』三島良績（切手趣味社 1964）
『切手集めの科学』三島良績（同文書院 1965）
『郵便切手製造の話』第4刷（印刷局朝陽会 1969）
『新版・切手と印刷』（印刷局朝陽会 1977）
『新版印刷事典』（大蔵省印刷局 1979）
『解説・戦後記念切手』Ⅱ～Ⅳ巻 内藤陽介（日本郵趣出版 2004～2006）
『解説・戦後記念切手別冊 年賀切手』内藤陽介（日本郵趣出版 2008）
▶日本郵趣協会発行
　『(JAPEX96) 記念出版　戦後の記念特殊切手1946-1955』(1996)
　『日本郵便印ハンドブック2008』(2007)
　『戦後記念切手1946-1971』記念・特殊切手研究会(2009)
　『(JAPEX11) 特別出品 国立・国定公園切手』記念・特殊切手研究会
　　(2011)
　『郵趣モノグラフ31 製造面から見た書状10円期の記念特殊切手』
　　永吉秀夫(2020)
　『切手画家 木村 勝の遺した資料 ‒戦後切手 1962～1984‒』
　　「木村勝の遺した資料」編集委員会(2014)

―カタログ Catalogues―
『日本切手専門カタログ 戦後編2010-11』(日本郵趣協会 2009)
『ビジュアル日本切手カタログ Vol.1記念切手編1894-2000』
　(日本郵趣協会 2012)
『さくら日本切手カタログ2025』(日本郵趣出版 2024)

―定期刊行物 Philatelic Magazines and Periodicals―
『郵趣』各号（日本郵趣協会）
『郵趣研究』各号（切手の博物館、日本郵趣協会）

―個人論文、記事 Thesis and Article―
「鯉のぼり切手の実用版は2つ」岡田芳朗（『早大切手研50年』
　早大切手研OB会 1999）
「るりかけす、うぐいす、らいちょう、ほおじろの定常変種」鈴木康弘
　（『Variation』合本第4集 バリエーション切手研究会 2001）
「年賀切手の定常変種（昭和24～31年）」粟篤吉（無料世界切手カタログ・
　スタンペディア『フィラテリストマガジン第7号』 2015）
「まりつき、初版と再版の分類」安藤源成（無料世界切手カタログ・
　スタンペディア『フィラテリストマガジン第40号』 2023）
▶『郵趣研究』掲載（切手の博物館 日本郵趣協会）
「国際児童福祉記念切手の定常変種と版の研究」
　青山宏（第4号 1992.冬）
「戦後消印別最遅発行記念・特殊切手」水谷行秀（第84号 2008-5）
「第2次国立公園切手のご当地使用」大久保幸夫（第89号 2009-4）
「高額記念切手の適正使用例を探す」神田明彦（第90号 2009-5）
「小型シートの切り抜き適正・適応使用例を探す」
　水谷行秀（第91号 2009-6）
「御所人形（三番叟）の普通シートの分類」横山裕三（第110号 2013-1）
「万国郵便連合加入75年記念5円のプレーティング」
　西正勝（第163号 2021-6）

『ビジュアル日専』ラインナップ　Publication Schedule of the "JSCA"

手彫切手編	小判・菊切手編	田沢切手編	昭和・新昭和切手編	産業図案・動植物国宝切手編	記念・特殊切手編 1894-1944	記念・特殊切手編 1952-1966（本書）
2023年3月刊行（既刊）	2020年10月刊行（既刊）	2023年12月刊行（既刊）	JSCA 1937-1948	2023年5月刊行	JSCA 1894-1944 Commemoratives	JSCA 1952-1966 Commemoratives
JSCA 1871-1876 Published in March 2023	JSCA 1876-1908 Published in October 2020	JSCA 1913-1937 Published in December 2023		JSCA 1948-1965 Published in May 2023		Published in October 2024

※図版はイメージです。採録の範囲や内容は変更となる場合があります。

ビジュアル日専　記念・特殊切手編1952-1966

2024年10月20日　第1版第1刷発行

発行
株式会社 日本郵趣出版
〒171-0031 東京都豊島区目白1-4-23
切手の博物館4階
TEL 03-5951-3416（編集部直通）

発売元
株式会社郵趣サービス社
〒168-8081（専用郵便番号）東京都杉並区上高井戸3-1-9
TEL 03-3304-0111（代）　FAX 03-3304-1770
【オンライン通販サイト】http://www.stamaga.net/
【外国切手専門ONLINE SHOP】https://stampmarket.biz/

編集・デザイン
最上邦昭　三浦久美子

印刷・製本
シナノ印刷株式会社

JSCA COMMEMORATIVES 1952-1966

Publication: Oct. 20, 2024

Publisher
Japan Philatelic Publications, Inc.
1-4-23 Mejiro, Toshima-ku, Tokyo, 171-0031
4th Floor, Philatelic Museum
E-mail: jpp@yushu.or.jp

Supplier
Japan Philatelic Co., Ltd.
3-1-9 Kamitakaido, Suginami-ku, Tokyo, 168-8081
【Online Shop for Japanese Stamps】http://www.stamaga.net/
【Online Shop for World Stamps】https://stampmarket.biz/

Editor and Designer
Mogami Kuniaki, Miura Kumiko

Printing
Shinano Co., Ltd.

令和6年9月24日　郵模第3096号
©Japan Philatelic Publications Inc. 2024
ISBN：978-4-88963-884-4

＊乱丁・落丁本が万一ございましたら、発売元宛にお送りください。送料は発売元負担でお取り替えいたします。
＊無断転載・複製・複写・インターネット上への掲載（SNS・ネットオークション含む）は、著作権法および発行元の権利の侵害となります。あらかじめ発行元まで許諾をお求めください。

趣味の切手 通信販売・買入

日本切手全般、格安品満載の販売カタログは、お電話いただき次第ご郵送申し上げます。

セキネ・スタンプ
〒110-8691 上野郵便局 私書箱112号
TEL：03-3872-6082　FAX：03-3872-6054

ミニオークション開催！

オークション誌は切手200円同封で郵送。
"WANT LIST" 受付けます！
TEL.090-1218-7717

ゼネラルスタンプ・品川
〒168-8691 杉並南局 私書箱15

出品無料！落札者の手数料無料！

「越中you趣」誌は令和創刊のオークション誌！
日本・中国新切手販売・会員寄贈品の無料頒布！
季刊で毎回出品数概ね400点以上！新規入会の方には本誌で利用できる「越中ポイント」300点進呈！直近の過去号を無料でお送りします。

＜お申し込み先＞
〒936-0059　滑川市晒屋337
越中趣味の会　日専係

切手買入

店頭販売：日本切手・外国切手

11時～17時 （月）（木）曜 定休
TEL.090-3815-5622
monalisastamp1127@gmail.com

切手の博物館1F
モナリザスタンプ est.1969
〒171-0031 東京都豊島区目白1-4-23

郵趣研究

オールカラー　**偶数月の20日発行**

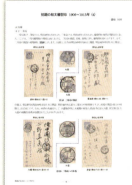

『郵趣研究』年間購読料（税・送料込）
- 紙　版：4,200円
- WEB版：3,600円
- 紙版＋WEB版：7,000円（特別割引価格）

本誌ご購読は下記からお選びください

① 振込
専用郵便振替口座にお振込みください。
口座番号：00160-6-3700
加入者名：公益財団法人日本郵趣協会

② 郵便払込用紙による送金
協会事務局にご連絡ください。
払込用紙をお送りします。

WEB版 郵趣研究

サンプル版をご覧ください！
スマートフォンで右のQRコードを読み込んでいただくと、WEB版郵趣研究サンプル版とご購読お申し込みをご案内します。

WEB版にはWEB版限定の特別付録があり、また紙版より2日ほど早く見られる特典があります。
発行：公益財団法人日本郵趣協会

1992年の創刊以来、さまざまな収集分野の研究・レポート記事を発信し続ける本誌は、オールカラー化で大変ご好評をいただいております。紙版より早く見られるお得な「WEB版」は、配信される貴重なマテリアル図版・写真が倍率を上げて見られ、保存資料としてもお役立ていただけるようになっています。

◇広告の中のJPSマークは公益財団法人日本郵趣協会維持会員です。

新料額普通切手12完

〈送料〉シート：600円　その他：110円

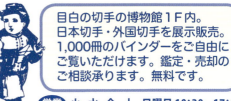

① 9月2日発売　新普通切手12完　1,460円
② 同上　100面完シート 額面売 (1,200枚)　132,900円
③ 同上　銘版入コーナー10枚群(120枚) (1割増)　14,620円
④ 同上　カラーマーク入10枚群(120枚) (1割増)　14,620円
⑤ Cartor新銘版1～500円11完10枚群(110枚)　14,500円

全て税込価格

〒150-8691 渋谷局私書箱80号　英国海外郵趣代理部
TEL：03-3499-5250

ロータスフィラテリックセンター

目白の切手の博物館1F内。
日本切手・外国切手を展示販売。
1,000冊のバインダーをご自由に
ご覧いただけます。鑑定・売却の
ご相談承ります。無料です。

営業　火・水・金・土・日曜日　10:30～17:00

〒171-0031 豊島区目白1-4-23 切手の博物館1F内
080-5514-6847・FAX 044-733-3388

開業28年の信頼と実績　宇中スタンプ JPS

中国切手 初日カバー、封筒、航空便を高額見積買取強化中！

長年の御愛顧誠にありがとうございます。中国切手を買取いたします。
買取値、相場影響、切手の状態等の不明点などのご相談は、
お気軽に買取相談専用電話へ。

☎ 080-3018-3695

書信館、蟠龍切手、清朝時代、中華民国初期等　旧中国切手、高額見積り買取を行っております。

- 誠実に、丁寧に詳しく、評価し買取いたします。
- お支払いは、ご郵送の場合見積り終了次第即日送金のいたします。ご持参の場合は、即金でお支払いいたします。
- 新中国切手はもちろん、旧中国、香港、台湾、そして、バラ、使用済み等すべて査定・買取を行っております。
- 全国無料出張いたします。お気軽にご連絡ください。

宇中スタンプ　〒167-0051 東京都杉並区荻窪3-47-21-601　E-mail：satohc03@yahoo.co.jp
TEL：03-6276-9668　FAX：03-6276-9669　携帯：080-3018-3695・080-4330-9988

郵便創始75年記念 初版

日本郵便切手商協同組合鑑定書付

小林スタンプ商会 JPS

〒121-0822 東京都足立区西竹の塚2-4-48
現金書留　ぱるる：10130-63623041
みずほ銀行足立支店(普)1369564　名義人「小林信博」

日本切手・外国切手専門店

NIHON PHILATELIC CENTER
Established 1950
日本フィラテリックセンター

ホームページ ▶ http://japanphilatelic.jp

コレクション作りのお手伝いから、継承・売却・仲介のご相談まで、よろず相談承ります。日本切手・外国切手コレクション即金買入致します。大口買取、出張買取、郵送買取、まずはお電話でご相談下さい。日曜、祝日も随時対応可能です。

オークション開催月(2025年より)：1月.3月.7月.10月
年4回オークション開催予定　フロア・メールオークション同時開催

年会費	2,000円 (2025年より改訂)	即売会への出店
見本誌	1部 500円 (切手代用可)	JAPEX・スタンプショウ等の併設ブースに出店予定です。(他の出店予定はホームページでご確認下さい。)

◆営業時間 10時～18時(日・月・祝定休)　◆〒530-0001 大阪市北区梅田1-2-2-200号　大阪駅前第2ビル2階
◆TEL (06)6341-8466 / FAX (06)6341-8480　◆ゆうちょ振替 00920-3-20839　◆nihon-philatelic@outlook.jp

◇ Phila は、JPSコミュニティ通貨「フィラ」取扱加盟店のマークです。